FIRE MEN

STORIES FROM
THREE GENERATIONS
OF A FIREFIGHTING FAMILY

GARY R. RYMAN

D1248698

TRIBUTE
BOOKS

First Edition 2011
Printed in the United States of America

Editor: Stephanie Longo
Cover design: Patrick Tigranian
Front cover photo: Rich Banick Photography, Scranton, Pa.
Interior design: Dave Justice
ISBN: 978-0-9822565-9-6
Library of Congress Control Number: 2011926994

Tribute Books
PO Box 95
Archbald, Pennsylvania 18403
(570) 876-2416
Email: info@tribute-books.com
Website: www.tribute-books.com

For information on bulk purchase discounts or for fundraising opportunities, please contact the special sales division of Tribute Books at: info@tribute-books.com.

Visit the book's web site at **www.Fire-Men-Book.com** and email Gary R. Ryman at gary@tribute-books.com.

All chapter photos: George Navarro
    Except:
    Chapters 1 & 20 photos: Michelle Ryman
    Chapter 8 photo: Michael Ryman
    Chapter 18 photo: Fred Bales
    Introduction & Epilogue photos: Gary R. Ryman

Publisher's Cataloging-in-Publication data
Ryman, Gary.
    Fire men : stories from three generations of a firefighting family / by Gary R. Ryman.
    p. cm.
    ISBN 978-0-9822565-9-6
1. Fire fighters—Biography. 2. Fire departments—New York (State)—New York—Officials and employees—Biography. 3. Rescue work—New York (State)—Biography. 4. Fire extinction—United States. 5. Fire departments—Pennsylvania—Officials and employees—Biography. 6. Rescue work—Pennsylvania—Biography. 7. Fathers and sons—Biography. I. Title.
TH9119 .R96 2011
363.3/7/0922–dc22   2011926994

To Michelle, Michael & Megan...

· ACKNOWLEDGMENTS ·

This book could not exist if it weren't for the time and dedication of hundreds of firefighters. The stories herein are based on my best recollections and I have attempted to tell them as accurately as possible. There is no doubt that some participants may recall some aspects differently; any errors in that regard are solely mine. Names, with very few exceptions, have been changed and the stories are mainly, but not entirely, chronological.

Every firefighter will notice some incidents in which accepted standards and practices for safety and use of personal protective equipment are violated. They are related in this way simply because that is how it happened and they are not to be taken as an example of proper firefighting technique; as my son, Mike, regularly tells me—"You can't do it that way anymore, Dad."

I would also like to thank the Endless Mountains Writers Group (EMWG) and, in particular, Hildy, Marcus, Jeanne, Carl, Rob, Ann, Eleanor, Mary, Mary and all the other group participants.

I am also grateful to George Navarro for his assistance in locating photographs. The placement of the pictures throughout the book is random. A photo does not necessarily depict an incident described in the corresponding text of a chapter.

• CONTENTS •

• INTRODUCTION •

A Wake-Up Call

At the age of twenty, I was a shit hot firefighter, convinced that I could walk through fire. My youthful delusions of immortality were instantly destroyed early one morning in March of 1982.

The whistle pierced the nighttime silence and the bunk room lights illuminated simultaneously. We tumbled from our beds, jammed our feet into our boots and pulled our bunker pants up, threw our suspenders over our shoulders, and snapped our buckles across our waists.

The dispatcher's voice came over the station speakers as we stumbled half asleep onto the apparatus floor. "Box 24-12 for the house on fire, 15721 New Hampshire Avenue. Engine companies 24, 4, and 15, Trucks 24 and 18, Ambulance 249, Ambulance 49 respond, 0611."

The apparatus doors began to open as we donned our bunker coats, nomex hoods, and leather helmets. I climbed into my pre-assigned spot in the jump seat of the ladder truck, which was right behind the officer, who sat next to the driver in the cab. The turbo-charged diesel motors of both the engine and ladder truck roared to life. Red and white lights danced, flashed, and rotated in the early morning darkness. With its one hundred foot ladder, the Seagrave tractor drawn aerial was sleek and elegant as a rocket ship, the full-sized version of the Tonka truck I had played with as a child.

The apparatus began to roll out the doors, first the engine and then the truck. As each made the left turn onto the avenue, the officer in the right front seat pushed a small orange button on the Motorola radio. A noise that sounded like a turkey gobbling squawked from the box and a yellow light flashed on and off, letting each officer know that the dispatcher, who was many miles away, had received the signal on his video terminal indicating that the vehicles were responding.

"Okay Engine 241, okay Truck 24, okay Engine 151, okay Engine 41, okay Truck 18," responded the dispatcher as he watched the signals on his terminal showing the change in each vehicle's status from "in quarters" to "responding."

Two beeps sounded over the radio speaker between the jump seats. "Engine Companies 24, 4, and 15, Trucks 24 and 18 responding, Ambulance 249, Ambulance 49 responding, Box 24-12 for the house on fire 15721 New Hampshire Avenue."

There was a tremendous amount of information contained in the few short radio transmissions made by this time. The fact that the dispatcher had told the units they were being sent to a "house on fire" rather than a "house fire" indicated to us that the dispatchers were confident there was an active or working fire at that particular location. The fact that they had assigned a full box alarm, which was typically reserved for apartment buildings or commercial structures, was also an indication that they thought a serious fire was in progress. Two engines and one truck would have been assigned for a typical house fire. The second transmission told everyone responding that, while all the fire apparatus were on the road, none of the ambulances assigned to the box had gotten out. Depending on what we encountered, that may or may not have been an important piece of information.

Secure in my jump seat on Truck 24, I strapped on my air pack, tightened the shoulder straps, slung the face piece around my neck, and connected the other end of the hose to the regulator on my waist belt. I put my gloves on, checked my flashlight, and settled back for the ride. Periodically, the siren screamed and the air horn blasted as we approached an intersection, but not continuously, since traffic was light in the early hours of the morning.

The engine was perhaps a quarter of a mile in front of us so I couldn't yet see the house we were going to, when, as expected, Sergeant K. announced over the radio in an exaggeratedly bored voice, "Engine 241's on the scene with a workin' fire. Have the next due engine take the hydrant at 15711."

"Engine 241 is on the scene with a working fire. Next due engine, take the hydrant at 15711 New Hampshire. Engine 151, are you okay?" the dispatcher repeated the officer's orders, making sure that the next due engine knew the assignment they had been given.

The engine had pulled to the rear of the house and, as we filled the driveway behind it with the long aerial, I could see the three inch supply line trailing from the rear of the hose bed back to the yellow fire hydrant a few hundred feet back on the street. I climbed down from my seat on the truck and noticed wires down in the backyard, sparking and jumping; something to stay away from.

The engine crew was stretching an inch and three quarter line from the cross-laid hose bed above the pump panel to the back door. Dark grey smoke was pushing out from the second floor of the house.

Looks like a room and contents, this should be a piece of cake.

Don, the tillerman who had been responsible for guiding the rear of the long aerial from the rear steering wheel on our early morning trip, and I pulled a twenty-four foot ladder off of the rear of the ladder truck, and raised it to the window of the room that was burning on the second floor. The truck officer, Pelcie, quickly climbed the ladder. When he thought that the engine crew had made the top of the stairs, he smashed out the window with an axe to ventilate the room and quickly returned to join us near the rear door.

Pelcie, Don, and I put on our face pieces and started up the stairs behind the engine crew to begin searching for any possible victims. The second floor was black; smoke banked down to the floor with zero visibility. Don went to the right following the hose line to check the room across from the fire room. Pelcie and I went to the left to search the bedroom adjacent to the stairs. We started to search, working quickly. Everything seemed to be going fine.

I worked my way around the room, going the opposite direction from Pelcie with my left hand on the wall to maintain orientation in the heavy smoke. I reached out with my other hand, sweeping the floor, furniture, the bed...any place where a victim might be found.

Suddenly, a huge blast furnace door opened. There were flames no more than six inches over my head, solid right to the ceiling. B.J., the nozzle man from the engine company, was lying in front of me on fire. He had neglected to put his hood on and pull his collar up before entering the building. The furnace-like temperatures ignited the exposed corduroy on his flat coat collar and the paint on his yellow helmet; the flames ringing his head like a burning halo. His screams were like nothing I had ever heard before. Even through his face piece, they were a simultaneous explosion of pain and terror.

I humped my body forward, trying to get to him while staying as low as possible under the flames. As I moved toward him, I accidentally smashed the exhalation valve on my face piece on the floor, popping a buckle loose on my mask, breaking the seal with my skin. I began to breathe smoke along with the compressed air from my tank. I was able to beat out most of the flames on B.J., but I still saw fire in front of me.

Where the hell was that coming from? I pulled my hands back toward me. Shit, my gloves were on fire. I beat them out on the floor.

I tried to get my seal back, pushing on my face piece, but to no avail. I instinctively reached for the bypass valve on the regulator. A firefighter is trained to know that when his mask malfunctions, he needs to turn the red valve, which allows air to flow directly from tank to face piece. *If I open the bypass and run out of air, I'm dead for sure; at least I have some air now.*

I looked out of the room toward the stairs for an escape route. Flames were going down the stairs, which isn't supposed to happen, but there they were. I looked for a window, but couldn't see one. My eyes were open to their physical limits; I tried desperately to see a way out through the smoke and flames. I was convinced that we were going to die; a fear-filled keening sound came from the back of my throat.

I hope this goes fast, 'cause it's going to hurt real bad. I hoped that I would pass out from the smoke before the flames got to me. It was the first time in my life that I truly knew what fear was. I was going to die here and there wasn't a goddamned thing I could do about it. I felt a combination of terror and resignation.

The sudden sound of smashing glass brought some hope of salvation. We weren't out of this yet, but maybe there was still a chance.

"Stay down and move toward me," Pelcie screamed, his voice muffled, but still loud through his face piece. I worked my way toward the sound of his voice, crawling to stay under the fire, half-pulling B.J. with me; I couldn't leave him there. I could barely make Pelcie out squatting next to the window he had just smashed, fire rolling over his head.

"Portable Truck 24 to Montgomery—EMERGENCY," he screamed into the microphone of his portable. "I need a ladder on side 4. I got a burnt firefighter!" He had heard B.J.'s screams as well.

The dispatcher quickly shut down all the other chatter on air and repeated the request. "Any unit on the fire ground, Portable Truck 24 needs a ladder on side 4."

A minute later, we heard a ladder slam against the right-hand side of the house; the tip at the base of the window. Don had somehow ended up in the room with us when the floor blew. Later on, the pump operator told us that when he turned around there was fire blowing out of every second floor window. Don went out the window first, followed by B.J.

Then it was my turn. I pulled myself up to the windowsill and leaned out. The normal air outside seemed frigid compared to the oven I was leaving. My mind was going a mile a minute. "MOVE," I told myself. "Get your leg up." As fast as my mind was working, my limbs felt like they weighed a hundred pounds apiece. They wouldn't do what they were told. The carbon monoxide I had been sucking in was starting to get to me.

Suddenly, I felt a hand on my collar and another on my ass, and out the window I went. I grabbed the sides of the ladder and began sliding head-first toward the ground. About halfway up or halfway down, depending on how you look at it, one of the engine guys caught me and guided me to the ground. The three of us who had just come out were herded to a spot under a large tree where the paramedics began to examine us. I was so damned happy just to be alive that I didn't give a shit about how screwed up I felt from the smoke.

The three of us and another guy, who had fallen down the stairs when the floor blew, were loaded into an ambulance and taken to the hospital. The ride was mostly a blur because of the high levels of carbon monoxide still coursing through my veins. When we arrived at the hospital, the first shift had just started. The nurses in their fresh white uniforms began stripping the bunker gear from us. Practically gagging from the smoke smell and soot, which soon ruined their clean dresses, they began to run various tests on us.

The doctor examining me checked the blood test results and looked down my throat. "Son, your throat looks like the inside of a chimney." He said I couldn't run calls for a day or so but after that I would be fine.

I asked to use a phone. They told me I could, but I would have to make a collect call. It was only eight a.m. by then; it just seemed like ten hours since we were dispatched. I dialed the operator and placed a collect call to my father at work. His secretary answered and when the operator announced the collect call from me, she started stuttering, "But he's behind closed doors."

I piped up and said, "Grace, it's Gary. I want to talk to him now!" About ten seconds later, he came on the line.

"What's the matter?" he demanded. He knew I didn't call collect just to say hi.

I explained what had happened and that I was in the hospital.

"You burned?" he asked.

"No, but I'm not sure why," I answered.

"Are you going to call home later?"

"Yeah," I responded.

"Just be real cool when you do," he instructed.

Dad only worked about a mile from the house and usually came home for lunch every day. I figured he would tell my mother about it then. At about five p.m. that day, when the long distance rates changed, I called them. That's when I learned that he hadn't said a word to her. He wasn't stupid. He knew how she'd react.

She started yelling, "Your father's been doing this for over twenty years and this never happened to him."

My "yes, but" answers weren't doing very well. I knew it was worry and concern on her part, but that wasn't making my explanation any easier. Over the next few phone calls, the volume went down but the butt-chewing continued.

What the hell had happened? It took a day or so until the fire marshals had finished their investigation and word came down to us. Dad wanted to know, too, and over the course of our phone calls—between my "yes, buts" to Mom—I shared it with him.

The house was an older style with a stairwell inside of a closet in the bedroom that had been burning, up to the attic. For some reason, a long time ago, tin had been put over the eave vents, making the attic tight.

The fire had extended up the stairs into the attic, but it didn't get too far. It was smoldering because of the lack of oxygen pushing back down the steps into the bedroom. This wasn't a problem while the room was burning, as the gases from the smoldering fire were consumed by the active fire.

Once B.J. had knocked the fire down with the hose and shut the nozzle, air rushed back in through the window we had taken out for ventilation. When that fresh oxygen hit the hot gases still pushing down from the attic—boom! A back draft occurred and we had fire through the entire second floor.

The night following that hospital visit, before we knew any of the hows or whys, I was lying in my bunk, mulling over the events of the day. My thoughts turned to an earlier fire, one of the first I had ever gone to.

• • •

I was about seven-years-old, playing with my toys on the kitchen floor and listening to bits and pieces of radio transmissions from a fire my father had been called to. There wasn't much to hear; it wasn't like today with practically everybody on the fire ground carrying a portable radio. It was the late 1960s. There may not even have been a portable radio anywhere on scene.

I could tell Mom wasn't happy. There was a sense of tension not normally there for most fires. Even as young as I was, I could feel it. I didn't understand anything from the radio transmissions, but she could and I knew she didn't like what she was hearing.

Families back then didn't have two cars so it wasn't like we could just drive over to the fire. The Dudley Court Apartments, the scene of the fire, were located on West Main Street in Endicott, New York. West Corners, Dad's department, had been called to assist.

Mom was on the telephone now, her tone hushed. I couldn't hear much of what she said, but I also wasn't paying a great deal of attention to what was going on. I continued to sense the tension, her movements were abrupt; she was scurrying from one side of the counter to the other. From the few comments I heard I could tell she was worried about Dad, far more than usual. I wasn't worried, though, because I knew my father was the best fireman around. At my age, fires were exciting, not dangerous, with big red trucks and sirens and cool coats and helmets. Even so, watching her move about, I got a funny feeling in my stomach about this fire.

Mom got off the phone and started gathering up coats and hats for my sister and me. She told us that Don, a family friend, was going to drive us over to see the fire. She tried to make it seem like a big adventure, but I understood there was a little more to it than that. I wasn't sure what, though.

Don picked us up and we drove over to the area of the fire. With all the fire trucks scattered around the building, their red lights flashing and blinking and hoses strewn in the streets like spaghetti, we had to park four or five blocks away from the scene. When we got there, I couldn't see any flames, but there was a lot of dark smoke coming from the apartment building and the air was filled with the odor of burning wood. There seemed to be a hundred guys in fire gear with long coats and three quarter boots pulled up to their thighs but I know there were far less than that. I remember watching Mom crane her neck, looking around. I knew she was looking for Dad. I looked for him as well, holding Mom's hand, her grip tight and moist with perspiration even on the cold day, standing among the crowd of spectators lined up along the sidewalk.

A few minutes went by but I'm sure it seemed a lot longer to her. I saw a guy with a long yellow coat and helmet dripping with water from the hose lines come out of the building, stop on the sidewalk, and drop to his knees to take off his mask. It was Dad! Just through the touch of her hand holding mine, I could feel a sense of relief pass through my mother, the tension flowing out of her body.

Later that day, I learned that two firemen had died in the fire. They were from Endicott, and had been wearing old-style filter masks, which didn't provide any air or oxygen to the wearer; useless in the viscous smoke of the apartment building.

Once Mom had seen Dad, she gathered us up to go home. I begged to stay and watch more, but that wasn't going to happen. More than a couple years went by before I fully understood the emotions of the situation. After my trip to the emergency room in Maryland, I wondered if my phone call took her back to that day in Endicott.

• • •

As for me, my next time inside on a fire didn't come for a few weeks. I knew I'd be nervous, but I had to get back on the horse. I looked forward to it and dreaded it at the same time. When that day finally came and I went through the door, my body was clenched so tight you couldn't have gotten a knitting needle up my ass with a hammer.

I learned I wasn't the shit hot firefighter I thought I was. *Hell, you can die in there.*

• CHAPTER ONE •

A Family of Firefighters

Ｈow did it all start? I am the second of three generations of firefighters. My father, Richard Ryman, joined the West Corners Fire Department near Endicott, N.Y. in 1962, when I was just a year old. He told my mother he would need to attend a weekly training session every Tuesday night for six months. Years later, I heard her joke many times that the six month training period was the longest one known to man, lasting as it did for some forty plus years.

Dad progressed through the ranks and ultimately became chief of the department. A job I think he liked better—and one he did for thirty years—was his position as a New York State fire instructor. You know you've done a job like that for a while when your classes hold the children and, in some cases, the grandchildren of firefighters you started with and taught years ago.

"Mr. Ryman," one of the kids would say, "My grandfather said to say hello to you." They would then relate some fire that both of them had been on back in the day.

I started on the ambulance at age sixteen, the youngest starting age possible in those days. In that area of upstate New York, Emergency Medical Services, or EMS, were separate from the fire service. It was a pretty busy outfit with three Advanced Life Support-equipped rigs running over three thousand calls annually. Back then it was all volunteer, twenty-four hours a day, seven days a week. The only reason I started on the medical side was that I couldn't join the fire department until I was eighteen-years-old. It gave me the opportunity to be involved in the emergency services business, and was certainly useful in the number of skills that I was able to develop.

Back in the old days there were no pagers and no remote radio receivers of any kind to alert the volunteers. Way back in West Corners, even before my father's time, when there were only a few named streets in town, a peg board with each street name written next to a hole hung on the station wall. When a fire call was received, the dispatcher in Endicott sitting at a Lily Tomlin-like switchboard would connect to a phone line, which would then activate the station siren for two and a half minutes. The first person to the station would answer the "red phone," which was a direct line back to the dispatcher. He—it was always a he in those days—would give them the address and nature of the fire call, and a peg would be put in the hole of the street that the fire was on for the next member's information. A later high-tech innovation was a chalkboard that the address was written on. This chalkboard was still in place well into the 1980s as a back-up. Primitive, but effective.

The system hadn't changed much from the peg board when Dad joined. He laughingly told me many times about his first call. In the middle of the night, he woke up to the siren wailing in the distance, down over the hill. He quickly got out of bed and dressed, racing to the car. He sped toward the station, less than a mile from the house, impressed with his reaction time and rapid response to the emergency call.

When he got to the station, he found he was a bit behind the curve. Numerous cars were already there, and all of the fire apparatus—two pumpers and a squad truck—were already gone. Luckily, the call was only right down the street; he could see the flashing lights at the nearby bank. Driving the short distance, he saw the apparatus positioned around the building and ground ladders raised to the roof. The fire was minor in nature, but he quickly figured out he needed to pick up the pace if he ever hoped to make it onto one of the fire trucks.

After that, Dad became an efficiency expert's dream. Clothes were carefully laid out on the bureau each night before bedtime. Keys, glasses, and cigarettes were strategically positioned. The most radical idea was yet to come: an automatic garage door opener. Those were unheard of in our neighborhood, but Dad took it to the next level. Most garage door openers, even today, have the button that activates them in the garage next to the car. That wasn't enough for Dad. He put an additional button in the closet in the bedroom which allowed him to hit the button while getting dressed. The garage door would already be open when he reached the garage, saving a good five seconds. A NASCAR pit crew would be impressed with his speed out of the house. When I was about eleven-years-old, we moved to a new house in a nearby neighborhood. One of the first things wired in was the activation button for the garage door opener in the closet of the master bedroom.

In the mid-1960s, a massive technological advancement happened—Plectrons became available. Plectrons were tone-activated radio receivers manufactured by the Plectron Corporation. As far as firemen were concerned, they were the greatest thing since sliced bread. Now they knew exactly where and what type of fire they were going to. The name Plectron for a tone-alerted receiver became the fire service equivalent of Xerox for copiers.

The original models weren't even solid state, instead they used tubes. The warmth from the tubes made them attractive to animals. My cat loved to sleep on top of the Plectron because of the heat it emitted. The cat loved it until the high pitched squealing tone alert went off at full volume. Then he would jump simultaneously up from the radio and off of the top of the refrigerator upon which it sat. It was a sight to behold.

Because of all this, as a young boy, the importance of speed out the door was ingrained in me. When relatives visited, I knew to advise them of safety measures I had developed out of necessity. If the tones went off, I would yell, "Quick, Grandma, get in a chair! He'll trample you." This came from the experience of being treated as a track hurdle while playing with toys on the floor when a fire call happened to come in.

One of the coolest things when I was little was to go with Dad to the fire station. The building was old, small, smelly, and cramped, but it had character. Heck, it was a fire station. It had been expanded a number of times as the community and the fire department grew. There were two wooden doors on the front of the building, but the bays were double in depth. When I was very small, one of the rear bays housed an old John Bean high pressure fog truck, a technology long out of vogue; the other a boat on a trailer. Shortly thereafter, the Bean was sold and replaced by a 1968 Dodge Power Wagon brush truck. There was a back drive through bay behind the main building, where the squad or rescue truck was housed. This was one of the early walk-in type rescue trucks with large removable spotlights on each side. It was fun to walk around in the rear box and look at the shelves full of exotic-looking equipment.

The fire trucks were the coolest jungle gym in the world. The deep American LaFrance burgundy color, the texture of the heavy diamond plate, the complicated levers and pressure gauges on the pump panel on the side, and the slight residual smoke smell on the hose carefully arranged in the rear bed atop the pumper were all fascinating to a young boy. I climbed all over everything I could reach. I knew from Dad's warnings what I could touch and what was off limits. One time, I was in the cab of one of the engines pretending to drive, the massive steering wheel in my tiny hands, as my feet dangled from the driver's seat. I looked down and noticed a silver round button on the floor. I didn't know what it was and, overcome by curiosity; I slid out of the seat and touched it with my foot. The screaming sound of the siren on the engine was magnified tenfold inside the building. I launched myself out of the cab, scared half to death. Dad wasn't upset. I think he figured I had learned my lesson when he saw how terrified I was. It was a week or two before I regained enough confidence to venture back into the cab of one of the engines, but I did go back, many times.

The firemen's room contained a massive, ornate pool table, a long wooden shuffleboard game along the wall, numerous couches and chairs, and a television. The far end of the room was occupied by the bar, behind which was the inevitable, at least in those days, beer tap. Coffee tables and magazine racks contained firematic-related magazines. Hidden back behind the bar was a pile of *Playboy* magazines. I knew these were off-limits to me, but I did peek once in a while when Dad wasn't looking.

The biggest treat for me was the rotating display of potato chip bags on the bar. If we were going to be there for a while sometimes Dad would buy me an orange soda and a bag of barbecue potato chips. Sitting on one of the butt-ugly orange padded chairs that were so common in the 1960s, I chomped on chips and swilled the sweet orange soda while looking through *Fire Engineering Magazine*. An unusual combination, but one I loved. It seems small, but it was a big deal to me.

Fires happen anytime, even on holidays. Dad's generation spent many hours away from home on pretty much any holiday you could think of. Christmas, Thanksgiving, Easter—name the holiday and I've probably gone out on a call. The pager can activate at any time. Besides holidays, nothing is exempt from interruption, including sleep, meals, showers, and even sex.

One Christmas Eve in the late 1980s with my wife Michelle's family filling our Pennsylvania home, a call came in to assist a neighboring department on a house fire. Away into the night I went, leaving the warmth and glow of the twinkling lights in the living room. Extinguishing the fire itself was routine; what wasn't routine were the remains of beautifully wrapped Christmas presents around the blackened tree, the now-melted ornaments still visible. My heart sank. While we weren't pulling the occupants from the charred debris, I could imagine the heartbreak of the family when they found the remains of their home upon returning from whatever gathering or celebration they were attending. Christmas was not as joyous for them, or me, that year.

Christmas was always special when I was growing up. We were always up before dawn to see what Santa had brought us. I truly believed in him. Once, I even thought I heard the hooves of reindeer tapping on the roof of our house while Santa was making his delivery.

One year, one of the presents I received was a multiple level gas station/parking garage in which you could drive and park and pretend to work on your Matchbox cars. My sister was receiving something called the "Imagination Dollhouse." Both had "some assembly required," somewhere in excess of a gazillion pieces.

My sister and I were sound asleep in bed, which we knew was important, because Santa wouldn't come if we were awake. Mom and Dad were getting out the hidden presents and the toys that needed assembly when the Grinch decided to pay a visit in the form of a house fire. The Plectron went off and so did my father, leaving Mom to finish putting the presents under the tree, and more importantly, begin the toy assembly.

Dad barely made it back before we woke up that Christmas morning. As usual, we were wide-eyed and thrilled with everything Santa had brought.

Years later, in the post-Santa period, Mom would regularly retell the story of that Christmas Eve, complete with uproarious laughter as she described the "millions of pieces necessary" to assemble the toys that year. She stayed up all night, the elf completing Santa's work.

• • •

The years went by and it was finally time for me to start in the family business. My first shift on the ambulance came on my sixteenth birthday, which was not your typical present. I had spent the preceding winter and spring months going through standard and advanced first aid classes as well as CPR training. Back then, sixteen-year-olds were not eligible to take the EMT course.

I spent the entire noon to six p.m. shift as nervous as a cat. I couldn't wait for the tones to go off so I could put my training into practice. I went through all the equipment on the rig multiple times to ensure I knew where every last item on board was located. I paced the floor from the bay back to the squad room a thousand times. Every time a set of tones started to sound over the radio I jumped, ready to go.

I'd love to say my first call came that shift, that it was a critical incident and an incredible experience. It didn't happen that way. My first call didn't come that shift or the one after that. It was probably four or five shifts before I got my first run, which I vaguely recall as a fairly routine house call. After that, it seemed like the calls never stopped.

Turning eighteen meant two things. I could now, legally, drink in a bar (the age limit not yet raised in New York) and, more importantly, join the fire department, which I immediately did. Living at home while completing an Associate's degree in Fire Science let me pick up some more experience and learn a great deal from my father.

In 1981, at age twenty, I enrolled in the University of Maryland to finish my Bachelor's. Instead of living in a dorm, I moved into a fire station in Montgomery County, Maryland, just outside Washington D.C. As a college live-in, bunker, or "wasted college fuck," the initial term of endearment used by the career guys, I started to get my real education.

The years that followed led me to the hills of Northeastern Pennsylvania, where I continued as a volunteer firefighter, eventually reaching the penultimate position of chief of department. I am employed as a fire protection engineer with a large commercial insurance company, which gave me the flexibility to work from home. More importantly, shortly after arriving in Pennsylvania, I met my beautiful wife, Michelle, on a blind date. After a courtship through which there were a number of missed or late dates due to fire calls, some of which I still hear about, we were married. A few short years later, Mike, the first of our two children, was born.

For his first thirteen years, it didn't seem like Mike had much interest in the fire department. Sure, he played with toy fire trucks and had a helmet, coat, and boots. He enjoyed going to the fire station with me and playing on the trucks, just as I had at his age. His interest, though, was not much different from any young boy's and, as he reached double digits in age, it seemed to wane. I didn't push it; I never wanted him to be a firefighter because of me, it had to be something he wanted to do himself.

When Mike turned thirteen, something happened and it was like flipping a light switch. He wanted to read the trade journals, look at fire-related websites and, more importantly, started asking to come along on calls.

I took him on a few and he would take pictures and sometimes help rack hose when things were done. He loved it; he couldn't wait until he turned fourteen and could join as a cadet. Now that my son is also a firefighter, that makes me the icing in the Oreo, the middle of three generations of firefighters and gives me the opportunity to see things not only as the son, but now as the father.

When I was young, I understood the horrendous damage a fire caused, and at the same time wanted to see more of it. As incongruous as this may seem, it is true and it is pretty much how Mike sees things through his youthful eyes. No one, save a psychotic, wants to see someone's property destroyed, much less injuries or fatalities. At the same time, practicing your craft is incredibly exciting and challenging, something you want to do as much as possible and therein lays the paradox of fighting fires.

As I've gotten older, my desire for action has waned and I've developed a more multifaceted understanding of the business. That doesn't mean the satisfaction, the bone-tired warmth from a good stop on a fire, goes away because it doesn't. Lying on the ground outside a still-smoking house, head soaking with sweat, swallowing a bottle of water in one long pull as steam still rises from soot-stained turnout gear remains a great feeling for me. Once the fire has been put out, I mentally review the events and then go back inside to examine what went right, what went wrong, and what could be done better. I make mistakes and find things that could have been done better on virtually every call.

The physical recovery after a fire went more quickly for me in my youth. My understanding was also more simplistic, less complete. The adrenaline rush and excitement of being inside with the fire used to be enough for me. Now, I appreciate the chess-like complexity of the fire, the situational awareness of reading the smoke and conditions and knowing where the fire is and where it's heading and how it could get there so as to be able to plan where and how to stop it.

At eighteen-years-old, I thought I was invincible; I'd have gone into a burning house of matches, having little understanding of the risk versus reward ratio so important in firefighting decision-making. "Risk little to save little, but risk much to save a lot," is easy to say but difficult in practice. The gloss of immortality lasted until I got hurt and recognized I definitely could die in there; that I hadn't was based partly on luck and partly on the training I had received and the skills of those surrounding me. Young firefighters either learn the hard way or through maturing with age.

In the beginning, I was so excited to go inside that the inherent fear was, for the most part, suppressed. Sure, there was a little tingling sensation in the back corner of my brain, but it was overwhelmed by the cool new experience. Now it's different. Fear is probably a bad word, as you can't go into a burning building on a regular basis if you're scared shitless. A reasonable degree of concern or apprehension is a good thing. If you're not the slightest bit scared going into a building being ravaged and eaten away by fire, I don't want to be in there with you.

· CHAPTER TWO ·

The Helmet

In the early 1960s, West Corners, New York, still retained some rural character and charm; even though it was slowly becoming more suburban, farms, fields, barns, and their associated denizens were still common.

Dad moved up the ranks from firefighter to lieutenant, captain, and then assistant chief fairly rapidly. The transition from rank to rank differs little today. There's a helmet from back then that hangs on the wall in my office. I look at it daily as I sit at my desk doing paperwork, paying bills, etc. It's an old Cairns from the late 1950s. There's nothing unusual about it, other than the friction loss tables, which give the pressure loss through hoses at different flows, taped inside; not something most guys, then or now, would do.

It's mainly a soot-stained white. Close observation reveals it wasn't always that color. It's not like many helmets today where the color is injection molded into the shell. You can see the yellow below through some chips in the white paint; and the black beneath the yellow. It has definitely seen a fair amount of fire.

Back in the day, when firefighters made the rank of lieutenant, they didn't get a new helmet. They kept their old one and painted it yellow. A new leather shield with the title would be attached to the front. Moving up to captain wouldn't change the color, but a new front piece would come.

When the owner made assistant chief, the helmet was repainted again, this time white. The owner wore it for a number of years while in that position until it was ultimately replaced with a "modern" helmet. Safer, more impact resistant, the new helmet was definitely an improvement over the old from a fire ground perspective. It didn't have the same character, though.

The old helmet, if you found it in a flea market today, would probably cost you five bucks. It's nothing special, except to me. Mike seems to appreciate it as well.

It was my father's helmet.

Even though, or maybe because Dad worked in a small town, there were a number of fires that required a little extra ingenuity. Take, for instance, a fire that happened at the local cider mill. The cider mill was a popular place for families to visit in the fall. You could watch the staff make candy apples and warm doughnuts along with fresh-pressed cider, the sweet smells emanating from each increasing your appetite by the second. Sometimes on a brisk fall Sunday afternoon, Mom and Dad would take us to watch the cider press at work; the golden liquid pouring from the bottom of the big old wooden apparatus. We would always take some cider home, but the big treat was a luscious candy apple, which I would start to eat immediately, the sugar coating stuck between my teeth.

Late one night, there was heavy fire showing from the block-long mill when the first units from Endicott arrived. They quickly requested help from West Corners. Dad was assistant chief at the time and responded to the scene. He met with the Endicott chief and was given the attack on the north side of the building.

He directed the incoming West Corners engine, a classic white-over-red 1960s vintage American Lafrance, to a hydrant off a nearby side street. This was long before pre-piped deck guns or lightweight portable monitors, the small one man units. The crew had to get the bulky balky deluge gun off the top of the engine. With the long steel barrel, and three hose inlets, it took three men and a boy to get it down.

The crew carried it toward the building, and Dad realized the best location for the gun was on the far side of the railroad tracks, which ran alongside the burning mill. This was also before the days of five hundred channel portable radios. In fact, he had no portable radio at all.

Looking around, he spied a police car idling in the roadway, its officer nowhere to be found. He trotted over, jumped in the front seat, and picked up the radio microphone. Realizing he didn't know which police car he was in, he got back out and checked the identification number on the front license plate.

Inside again, Dad called the dispatcher over the radio and explained why he was on the police frequency and that he needed them to call the railroad and get the tracks shut down. A train running over three two and a half inch lines feeding the deluge gun wouldn't have done the hose, or the train, any good.

A few minutes went by and the dispatcher told him the message had been delivered. He went back and signaled the crew to begin stretching the lines across the tracks.

It was a long night but, working from both sides, they saved the rear half of the building which, after some repairs, continued to operate.

Over thirty years later, we still occasionally get cider, doughnuts, and candy apples from the same cider mill. Every time I take a sip, I think of Dad's story.

•   •   •

Fires weren't the only interesting calls back then. Late one night, or early one morning, depending on your perspective, the fire department was dispatched for a car accident on Route 17C. Dad arrived on scene to find a number of drunken kids along the roadway. It was pitch black on the two lane highway, both sides heavily wooded. The vehicle they had been riding in was down over the bank.

Dad questioned them about how many had been in the car, but it was yielding less than fruitful results, to put it kindly. These kids were so smashed they didn't know how many people were standing in front of them, much less how many had been in the damned car.

With no reliable answers forthcoming, it was time for a trip down the bank. It wasn't an enjoyable one in the dark, through the bushes, unseen holes, and heavy brush. On his way down, Dad noticed a tree clipped off fifteen feet above the ground.

If there was anybody left in this thing, it wasn't going to be pretty. He finally reached the car, which was on its roof in the soggy underbrush at the bottom of the bank. Dad shined his flashlight around, looking in the windows on one side of the car. No one there. He sighed with relief. But when he reached the opposite side of the car, he found he'd been mistaken.

There were two feet sticking out from under the roof. Amazingly, they were moving.

How the hell could this kid be alive?

Rodney, the chief, had arrived on scene and was hollering down to Dad.

"Dick, Dick, is it bad?"

Dad knew Rodney had a very weak stomach, and adding the contents of his belly to the scene in the woods was not going to be helpful at all.

"Stay up there, Rod," he answered.

The ambulance crew had worked their way down by now, as well as more firefighters. This was long before fancy airbags and the hydraulic rescue tools we have today. Ingenuity was the best thing they had.

A wrecker backed up to the spot of the now-missing guard rail. The cable was run out and draped over a large limb of a tree near the car. Chains were attached to four points of the undercarriage and then to the hook on the cable. The wrecker took up the cable and slowly lifted the car off of the kid. The ambulance crew slid him out and onto a backboard.

The trip up the bank was not a lot of fun but at least there were plenty of folks to help by that point. The kid made it. He had been very relaxed, so to speak, and the ground had been soft.

The state troopers had a few more details by this time. These kids had been at a new type of event, a garbage can party. The concept was simple. A new, clean galvanized garbage can was purchased. Everyone attending had to bring some kind of alcohol. Beer, whiskey, gin, wine, bourbon; anything at all was acceptable. All the booze was dumped into the garbage can, becoming a high proof punch of sorts, which the attendees spent the evening consuming. It's a wonder they tried to drive home; they could hardly stand up, much less walk.

Dad's stories were always great, but it was much more fun when I actually got to go to the fire. Those opportunities were few and far between when I was a kid. One summer Saturday when I was about eleven-years-old, a call came in for a house fire not too far away.

I ran down the hill from our house, following Dad's car. The house on fire was two blocks down, close enough to get there quickly and maybe see a good blaze. There weren't many fires in our neighborhood, so this was a rare chance. Turning the corner at full speed, my Keds practically smoking, I could see Dad's car sitting on the side of the street, with the red and white rotating light flashing away. Expecting to see lots of flames and smoke, I was disappointed as there was only a light mist of smoke coming from the second floor.

It was all over with and I didn't even know it. The homeowner had directed Dad to the second floor when he pulled up. Reaching the landing, he found the curtains burning in one of the bedrooms. Ripping them down, he gave them a quick stomp and then, still partially flaming, flung them from the open window.

Second thoughts set in as soon as he had done it. He knew he probably should have looked first.

Dreading what he might see, he stuck his head out the window to see where, and more importantly, on what, the burning curtains had landed. He then let out a sigh of relief. They had landed on the driveway, just missing a parked car. Rodney, the chief, had arrived and was stomping out the remnants with his shoes. He grinned up at Dad, knowing what he was thinking. They were both happy it was a minor fire. Everybody was but me. I walked back up the hill, disappointed I hadn't gotten to see a good one.

Fires and accidents weren't the only emergencies handled, then or now. In June of 1972, Hurricane Agnes wreaked havoc throughout large portions of the northeast, including little old West Corners. Dad and the rest of the department worked for well over twenty-four hours straight during and following the storm rescuing trapped people, pumping flooded cellars, and various other incidents.

The department had what looked like a pretty decent-sized boat; at least it looked decent before Agnes came. It was a V-hull with three rows of seats and an outboard motor. There was only one pond of any size in town and, on a normal day, the boat wouldn't even float in Nanticoke Creek, which flowed through town. The boat was plenty big to serve the community.

When Agnes arrived, Nanticoke Creek didn't stay small for long. It raged over its banks and flooded a big swath of the surrounding area, including a nearby trailer park, trapping a number of folks who had ignored evacuation orders.

Dad and the crew got the boat ready to launch. There was no problem floating it now; plenty of water was available. Not a strong swimmer, he was nervous, but he had a job to do. They got the boat in the water and started upstream. That's when the trouble began. The boat and its motor were no match for the now rampaging creek. At full throttle, the boat would do little more than stand still. The boat would go cross stream, if angled properly to handle the current, so they adjusted.

There were no swift water rescue teams or Gumby suits then. Thick hemp ropes, hip boots, and waders were the primary tools to get these people out. Going cross current, they walked in to the stranded folks by holding onto the side of the boat operating cross current. Not a textbook maneuver, but it worked. Nobody died, but there were a whole lot of very tired firemen after that.

The boat? Well, shortly after the flood was over, a "for sale" sign went up on it.

To say that Dad could be a little bit anal about equipment organization would be putting it mildly. I think it was the ex-Marine in him coming out.

The engines varied in vintage from 1957 to 1975 back when he was a chief in the 1970s and 1980s. What didn't vary is where things were located. You could open any compartment on any of the four engines (three first line and one reserve) and each piece of hardware, nozzle, appliance, wye, gate valve, etc. would be found in exactly the same spot on every piece.

Hose was a pet peeve of his. We had a spare load of hose for each engine stored in doughnut rolls on hose racks in the rear of the building. I would catch him regularly rearranging the hose on the racks so the end butt of each roll was in perfect alignment.

If he saw you put a roll of hose on the racks and not line up the butt with the adjacent ones, you would hear about it instantly. This was not one of his saner practices.

Dad was terrible with names. Guys in the department upwards of five years were, "Hey, you." If he did know your name in less time, it was not necessarily a good thing as there was likely a bad reason why he remembered it. At least when I joined, he had no excuse not to know my name.

One difference when you're the son of a chief—and Mike has had this experience as well—is that there is not always a realistic expectation level for you. I couldn't just do something as well as the next guy; I had to be better. The often asked question was— "Is he as good as his old man?" I'd like to think the answer, in both of our cases, is yes. I enjoyed the hell out of the few years we got to fight fire together. I hope Mike ultimately feels the same way.

• CHAPTER THREE •

My First Fatal

Starting on the ambulance at sixteen-years-old brought the reality and gravity of emergencies home for me. It also gave me the chance to start working with my father on some calls as well.

Just around dawn on a late spring day in 1977, the Plectron went off for a car accident with entrapment at Glendale Drive and Route 17C in West Corners. Since the ambulance was also due, I went along.

The car smashed straight through the guard rail at the T intersection and landed down over the bank into the woods after rolling through the brush. I scrambled down the bank to the car along with the arriving ambulance crew. The little Triumph TR-7 was a pile of mangled sheet metal, but we could see that the female driver was alive and conscious through the narrowed window beneath the crushed roof.

Before the widespread availability and use of the famous "jaws of life," extrication methods relied more on hand tools and a few more primitive methods. The squad arrived, and tools and equipment were passed over the bank to an area near the car. Dad and the ambulance crew chief coordinated how the EMS folks wanted to take the patient out and how the firefighters would open up the car. I took it all in—a young pup learning the ropes.

The two of them decided that the roof needed to be removed. Someone would need to be in the car with the victim while this operation was conducted to keep her calm. Back then, I was a skinny shit and the smallest one there, which made the choice for them. I leaned through the distorted window opening on the passenger side, and two firefighters picked my legs up, and slid me into the car. They handed me a blanket, and I covered the victim and myself in a makeshift tent, while trying to explain to her what was going to happen. With the strong odor of marijuana in the car, I wasn't sure how much she understood.

As the firefighters prepared to remove the roof, a protective hose line was advanced over the bank. I heard the infamous K12 gas powered circular saw start, and then the noise level in the car redlined as the saw bit into the steel posts. Sparks flew all around us, but we were relatively safe under the blanket. A couple of loud minutes later, the roof was lifted off the car and the blanket removed.

We immobilized the victim on a short backboard, and gently lifted her from the car, strapping her on a long board for the trip back up the bank to the waiting ambulance. Ropes and lots of manpower aided her trip up.

The old time methods may not have been pretty or quiet, but they did work. It was a very interesting first "pin job" for a sixteen-year-old. The victim had a few broken ribs and one broken leg but made a complete recovery. The results weren't always that good, though.

• • •

One weekend morning, I heard a call come in over the scanner for a motor vehicle accident at Route 26 and Carl Street, and begged my father to take me down to it. In those days, the fire department was not automatically dispatched on automobile accidents. Initially reluctant, he relented, and we drove over, much too slowly for me. Upon arrival, I saw a car with the front end smashed and wrapped around a telephone pole. Walking over, I saw a burlap bag over the driver's window. A man, a passerby it seemed, stood by the side of the car retching. As I approached the car, he said, "You don't want to look."

I know that, but I have to. It's my job now.

I pulled aside the burlap to check the driver. His head was thrown back and there was a jagged laceration across his throat. He definitely wasn't breathing. I reached in, grabbed his wrist, and checked his radial pulse. Ordinarily, as I had been taught, I would have checked the carotid pulse, but there was no way I could put my hands into that bloody mess, and determine anything. No pulse, no breathing; dead as far as I could tell. My first fatality.

The ambulance then pulled up. Hubie, the captain, got out and walked up to me. I waited for him near the driver's door.

"He's gone," I told Hubie. His raised eyebrows and the look on his face said it all. Hubie flipped back the burlap to check the victim himself. *Boy, I hope this guy really is dead, because if he isn't, I'm in deep trouble here.*

He was and I wasn't; in trouble that is.

It was a somewhat in-your-face experience for a young kid. Sure, I'd seen dead bodies in funeral homes, but it's different, especially the first few times, out in the real world. It didn't really bother me; I knew it was part of the job. I was so interested in the wounds and injuries that understanding that this man had a family who would be distraught didn't sink in until much later.

• • •

In a town as small as West Corners, it was inevitable that I'd run into someone I knew on a call but there really was and is no way to prepare for it. In the middle of the night, my ambulance Plectron had gone off with a report of a serious auto accident on Day Hollow Road. "Car over the embankment," the dispatcher announced. I ran down the hall and woke Dad up, telling him it was likely the fire department would be needed, and he agreed we should start out.

We got on scene around the same time as the ambulance. The good news was that there was a driveway available to take us to the bottom of the bank where the victim was located. We drove in and didn't have to worry about getting the victim up to the main road over the wooded embankment. The bad news was the trip over the bank cut the car almost in half, ejecting the victim; who was now in critical condition.

The victim lay on his back at the bottom of the tree-strewn bank. There were numerous head and facial lacerations. He was "doing the funky chicken," a description we had for involuntary movements of the arms and legs, likely symptomatic of brain injury. His respirations were irregular, or Cheyne-Stokes, where his breathing waned interspersed with scary periods of no breathing at all.

The on-duty crew and I went to work on him, and there was plenty for all of us to do. We cleared his face of blood as much as possible, inserted an airway, and supported his breathing with a bag mask and oxygen. We still had no idea who the victim was; probably a good thing at this point.

Dad held a couple of flashlights to give us some light. The medic began starting two IV lines, one in each arm, while we prepared to immobilize him on a backboard. We got him packaged and into the rig and the crew transported him to the hospital. Based on his condition, his prognosis was not encouraging. Dad and I went home and, after washing up, I went back to bed, having no idea who the stranger was we had just treated.

The next morning I got up hours after my father did, like any normal teenager. As soon as I came downstairs he broke the news to me. It was my friend, Dan. How he had found out, I don't know. We had lived next door to Dan and his brother, Donnie, on Grant Street for at least seven years. Donnie and I were the same age, and Dan a couple of years older. The three of us spent endless days together playing wiffle ball, football, and going sleigh riding in the winter. It took awhile for it to sink in.

We never found out who was actually driving that night. There had been another guy in the car, but he had walked away from the wreck. It came as no surprise that he told the police that he had been the passenger.

A few days later, Dan's parents made the heart-wrenching decision to remove him from life support. I'm told it was at least some comfort to them that people he knew had been there with him that night and we could assure them he had not been in any pain.

The funeral was hard. In a small town, things sometimes hit close to home.

• • •

West Corners was only a small part of a large area covered by the ambulance. I was therefore able to work with other fire departments. There were five other fire departments, two career and three volunteer in the Town of Union in upstate New York, and we ran calls with all of them.

The baby shit green El Camino had crashed into the pole, the front end demolished. As I looked inside the car, I could see the driver lying on the passenger side, his head between the seat and door jam, the remainder of his body angling down and away beneath the crumpled dashboard.

I had only recently been promoted to first attendant, or crew chief, and I was still just sixteen-years-old. The captain had moved up a few of us younger guys who were willing to assume more responsibility, and who he figured wouldn't kill anyone. I knew I was in over my head that day, but I had to take charge. I was scared shitless and fear, sometimes, is a hell of a motivator.

There was a massively wide laceration across the victim's forehead. The blood ran like a river down his face and into his mouth. He was literally drowning. The crew brought me an airway and the suction unit. Inserting the catheter in his mouth, I filled the suction reservoir to overflowing and then inserted the airway. It was enough to keep him breathing.

With this crisis out of the way, I continued my initial assessment. I ran my hands around the back of his head. My fingers found another massive laceration and, in following the contour of his head, they were into his brain before I knew it. Not a good sign. I continued down his body. His chest was okay but reaching further in the darkness to the abdomen, I found my hands filled with his intestines. He was eviscerated. I couldn't reach any further, but I knew his legs had to be broken as well from the angle of his body. That was the least of both of our problems at this point.

I had called for the fire department but we started working on the door before they arrived. A neighbor ran out of his house with a crowbar, and the three of us on the crew, adrenaline racing through our bloodstreams, practically ripped the door open. Engine 2 from Endicott arrived and their crew helped us bend it back, increasing the opening, as we held his head in place.

We slipped a backboard under him and as quickly and gently as we could, slid him out of the car. We had to reduce the one compound leg fracture, not a typically recommended technique in the field, but getting that bone out of the way was the only way to get the leg past the dashboard. We didn't screw around splinting anything; we needed to get him out and moving. In the back of the ambulance, we covered the evisceration with trauma dressing and poured sterile water over it to keep his intestines moist. We got some high-flow oxygen on him and made sure his airway was still clear. The medic who had arrived unsuccessfully tried a blind stick to get an IV established; since he was so deep in shock we couldn't even see any veins.

We took off for Ideal Hospital on the other side of Endicott at what seemed like warp speed, and at the same time it felt like it took forever to get there. I hit code 033 on the touchtone pad attached to the radio and called the hospital. "Union 105 to Ideal Emergency..." I gave the ER nurse a rundown on the victim's injuries starting with the most severe. The list seemed endless. The vital signs I reported were not encouraging. Everything I had to report took so long that it seemed like we pulled into the hospital ten seconds after I hung up the microphone.

The driver backed the rig up to the ramp. The ER crew had actually come out to meet us, which was highly unusual. The rear doors of the ambulance were jerked open, and Susan, one of the ER nurses, appeared in the opening. "Is he dead yet?" was her first question.

Hours later, he was stable but critical, and a few weeks later he actually walked out of the hospital. How, I'll never know. The cops told us later he had been drag racing when the wreck occurred.

It took us hours to clean the blood out of the rear of the rig. AIDS and other blood-borne diseases weren't in the picture yet, so gloves were almost never worn. It wouldn't have mattered; we were covered in red to our elbows, and it took quite some time to get ourselves cleaned up as well.

Overall, it was quite a challenge for a young kid and crew. We were pretty happy with the results.

● ● ●

It took a while after I got my driver's license before Dad would let me take the car on a call, concerned that the excitement might lead to excessive speed. I've gone through the same process with Mike, but all it took for Dad to acquiesce and trust me with the car was a single incident.

On this day, a mother, father, and their son were laid out on the dew covered grass of their front yard on Nanticoke Drive in Union Center, CPR in progress on all three. The beautiful weekend morning had been interrupted in a most ugly way.

I heard the call come in and the subsequent frantic calls for assistance over the scanner. I begged loudly enough that Dad agreed to take me. I had just gotten my driver's license, but he didn't want me taking the car by myself, concerned about my right foot getting too heavy in this situation. Once there, I trotted up and went to work on the victims, while Dad tried to make himself useful by helping with traffic control.

Kevin, one of the medics, knelt down beside me where we were working on the mother. He had a Life Pak defibrillator in his hands. Yanking the paddles free from the case and slapping them onto her chest, he directed me to "read it." The bright sun made it almost impossible to see the monitor screen. I cupped my hands around it and peered between them.

"V-Fib," I called out. "Hit her!" I pushed the button on the console to charge the unit. The tone sounded, indicating that we were fully charged. "Clear!" Kevin called, and we all moved to break contact with the victim. Wham! The electrical charge from the unit shot through the paddles and into the victim. Her body jumped slightly into the air.

Unfortunately, with Kevin on his knees beside her on the wet grass, the current shot into him as well. The paddles dropped from his hands as he flew backward onto his ass. He quickly recovered, shaking his tingling hands. He slapped the paddles back onto her for another look. No change. Continue CPR.

We moved to the next victim, the father, and repeated the sequence. Same result, except that Kevin didn't end up on his rear end this time. On to the kid. Same sequence but a better result. We got a rhythm. Multiple responders and medics were working by now. IV lines were being established, drugs pushed and repeat defibrillations performed.

The family had been using some sort of acid solution to clean their septic tank. Built-up fumes caused them to collapse and ultimately go into respiratory and cardiac arrest. Two injured firefighters were also being treated.

The first guy was not in great shape. He jumped in to try to make a rescue, but, not thinking, he didn't wear a breathing apparatus. He quickly became the fourth victim. The second firefighter on scene, smart enough to don an air pack, got everyone—including firefighter number one—out of the tank. Unfortunately, he had a leak on his mask seal and ended up taking in some fumes. He was conscious and alert and only in need of some oxygen and monitoring.

The normal EMS routine and protocols at that time weren't designed for an incident like this. After the initial IV, the medic typically would receive a verbal order over the radio from the hospital for each drug pushed. With three full arrests, we didn't have time for that. Standard protocols were being followed as far as the type and dosage size of the drugs pushed, but we didn't slow down to get the actual orders. Typically only medics pushed drugs, but that went out the window as well. I was pushing boluses of sodium bicarbonate, epinephrine and whatever else the medics needed. The syringe was slapped into my hand, and I'd insert the needle into the IV line and press the plunger at the rear, injecting the drug into the stream.

The patients were packaged for transport one by one. By the time we were ready for the last one to go, it became apparent that we were one ambulance short. We had two choices: wait for a rig to come back from the hospital or transport in Union Center's step van squad, an old bread truck-type vehicle that carried their rescue equipment. Since we figured it would take at least fifteen minutes to get a rig back from the hospital, and longer to get another ambulance from a neighboring jurisdiction, the choice was easy. The victim was strapped on a back board and laid on the floor of the squad, surrounded by the crew picked to transport. The Union Center chief jumped in the cab, and away the old bread truck went.

The kid made it but his parents did not. Both firefighters recovered from their injuries. Our folks who had done mouth-to-mouth had some burning on their lips, believed to be from the acid used in the tank.

The whole incident was quite a learning experience for many of us. I know it was for Dad. The next time something bad came in, he just gave me the keys and let me go.

• CHAPTER FOUR •

EMS Days

Emergency rooms save people every day but that doesn't mean mistakes don't happen. You learn this quickly on the EMS side of the business.

One day we brought in a possible heart attack victim. He was unconscious, but had a rhythm. After we left him in the treatment room still very much alive, we went to restock the drug box, replacing the items we had used on the call. Walking back toward the front desk, I passed the treatment room again, and saw the staff doing CPR on our victim. I watched for a minute or two and then noticed one of the leads to the heart monitor had become disconnected from the pasty on his chest. Walking into the room, I tapped one of the nurses on the shoulder and pointed out the discrepancy.

"Oops," the nurse responded as she reconnected the lead. They stopped compressions to check the rhythm and he did indeed have one. One of the first things they teach you when becoming an EMT is that doing compressions on someone who has a heartbeat is a bad thing.

Another time, we brought in a victim in full arrest with an esophageal airway, which we had put in to help support the respirations. This was pretty new equipment at the time in the late 1970s, and it had been introduced as a much easier tool to use rather than intubation—which they hadn't given our medics permission to do yet. The tube went in following the curve of the throat, and a balloon inflated with air, holding it in place. Air holes were in the tube right near the epiglottis, providing a good direct air flow to the lungs from a bag mask or similar appliance.

The ER doctor wanted the airway out so he could intubate the guy, but didn't know the protocol for removing the esophageal airway. He wouldn't listen to a thing as we tried to explain to him about the airway, which he'd never seen before. Instead, he just yanked it out with the balloon at the tip still inflated. Then he wondered why blood gushed from the victim's mouth. We just walked out, disgusted with the doctor. None of us wanted to see any more, including the ultimate pronouncement of death, which was inevitable after what the asshole did.

As in most things, one bad experience can overshadow ten good ones. Fortunately, the good guys outnumber the bad by quite a few, the jerks being more than balanced out by some exceptional doctors. Dr. June worked the eleven to seven night shift at one of the local hospitals. At the time, I thought she was middle-aged, but she couldn't have been more than forty but when you're sixteen, forty is middle aged. She was unpretentious and joked regularly with the young ambulance crews bringing her the local carnage on a nightly basis. She had a serious side as well, and would spend hours answering our technical questions.

Dr. S. was a young guy just completing his residency. He insisted you call him by his first name, Joe, when patients weren't around. Just that small gesture made him a crew favorite.

Dr. K. was a general practitioner. He put in night and weekend shifts working the ER, whether for the extra cash or to get away from his wife or for whatever reason. One slow evening, he spent hours working with me on reading EKGs and teaching me the subtle difference between first and second degree block. He was one of the doctors who listened.

He was working one night when another patient came in with the still new esophageal airway in place. Like the asshole doctor, he had never seen one before and had no idea of the protocol for removing it. Unlike the asshole, he wasn't afraid to admit he didn't know something and asked. The nurses didn't know either. I volunteered the information we had been given in the training class on how to remove one. He listened carefully, and realizing, at the very least, that it wouldn't kill the guy, told the nurse, "Sounds good to me. Set it up and let's do it that way."

After my EMT training, I took the classes necessary to get to the next level, which, at the time, was IV Technician or in fancy terms, AEMT II. As part of this, we had to spend time in an ER to start IVs under the supervision of the staff before we could do it in the field on our own. Following the classroom instruction, we had to get ten sticks in the ER before we were considered certified. Unfortunately, we were at the mercy of the flow of patients. Some nights, we could get two or three; at other times, entire shifts would go by without anyone coming in needing a line.

My first stick was a middle-aged woman who came in feeling very ill. She could not remotely understand what was wrong with her. That evening, she had attended a wedding reception and had a few too many drinks. Apparently, it had been more than a couple years since she had been drunk and she didn't recognize the symptoms. The ER doctor did, though, but no amount of explaining could convince her that she wasn't deathly ill.

"Go ahead," he told me. "Start a line with D5W." Sugar water. It wouldn't hurt her and made her think she was receiving medication to cure her ills. Sure enough, after a full bag, and more importantly a couple hours for her body to process the booze, she felt much better. Medicine is a wonderful thing.

The small, old-fashioned ER at Ideal Hospital where I did my time was a lot of fun. Ideal is long-closed now; absorbed by a large conglomerate like many small municipal hospitals and shut down as the bigger facilities are considered more efficient.

The emergency room staff consisted of a doctor, two nurses, an aide, and a receptionist. When you had four or five patients at one time, which is nothing for the large hospitals, it was nowhere near enough. At such times we really got to pitch in and learn a lot. It wasn't like some of the other hospitals where you were simply an observer until it was time to start an IV, returning thereafter to observer status.

At Ideal, they gave you as much responsibility as you could handle. Hell, you were free labor for them.

One Sunday evening I was going to go and put in a few hours after eating dinner. Just as we were getting ready to sit down, the phone rang. It was one of the nurses.

"Are you still going to come in?" she asked.

"Yeah, as soon as I eat."

"Can you come now?"

This was getting really weird. "Sure, but why?" I asked.

"We've got a gunshot wound, shotgun to the stomach, coming in from near the Pennsylvania line," she explained. They weren't patients or people; we referred to folks by their ailment: gunshot wound, head injury, appendix, etc.

"I'll be right there," I told her.

I got to the ER a few minutes before the gunshot victim. It was an all-hands evolution working on him. Everybody, including the receptionist, pitched in. I ended up working as the scrub nurse for the doctor putting in a cut down line in his subclavian. This surgical procedure put a large bore needle into a big blood vessel under the collarbone to move a lot of fluid quickly. Not what you'd get at most hospitals, which would frown upon an eighteen-year-old kid retracting the incision while simultaneously handing the doctor instruments.

The best part of Ideal Hospital, I have to admit, were the cute young nurses working the least desirable three to eleven shift. I never had much luck with any of them, but at least the scenery was pleasant. The bigger hospitals may be more "efficient," but that doesn't necessarily mean better care. And for a young EMT, they're nowhere near as much fun.

• CHAPTER FIVE •

Ambulance Runs Aren't Always Sad

Running ambulance calls wasn't all blood and tragedy. We had our share of fun as well. Round Top was a small park at the top of a hill on the south side of town overlooking the Susquehanna River. It was a well-known parking area or lover's lane-type place.

Occasionally on an eleven to seven shift we would cruise up that way in the ambulance to see if there were any couples making out. If there were, we would kill the engine and coast into the parking area in neutral, headlights off. Once close, we'd flip all the red lights on. Cars would rock, clothes would fly, and we'd laugh like hell, kill the lights, restart the engine and go on our merry way before they could figure out that it wasn't the police.

The group of us that typically covered the Saturday eleven to seven shift in the late 1970s were all young guys, growing boys with healthy appetites. Down the block from our squad room was a bar called Kelly's that also made fantastic pizza. Occasionally, someone would place a takeout order and fail to pick it up. When that happened, our phone would ring. Upon answering, a deep guttural voice would immediately begin talking with no pleasantries or discussion. "It's Kelly. Got a pie for you. Come get it." We'd jump in the rig and run down and he would give us whatever pizza had not been picked up. You never knew what it would be, what toppings, etc. But we got it for free. You couldn't beat it.

Along with the Saturday eleven to seven shift, we would regularly take the seven to noon shift on Sunday morning so we could sleep in if we didn't get a run. When we finally got up at nine thirty or ten a.m., we'd take the ambulance and go to breakfast at a nearby Friendly's restaurant where they liked us. We'd take our time and enjoy a nice leisurely breakfast. There was an ulterior motive to this. If we got a run while eating breakfast, obviously we would have to leave. When we returned, they would give us a new breakfast, but we'd only be charged for one. We didn't get the two for one every week, since we couldn't predict our calls, but we got it often enough that we made sure we were regulars there. Cops like donut shops. We liked just about anything.

* * *

We didn't always enjoy food, though. The elderly lady was complaining of abdominal pain. We had no idea what, specifically, was wrong with her as we loaded her stretcher-bound form into the back of the ambulance.

It was a small rig, a converted Chevrolet Suburban, common before vans and modulars took over. Two of us in the back along with the patient made for tight quarters.

She didn't know what ailed her either, but we all soon found out. There was an intestinal obstruction. The diagnosis was easy. Just as we got ready to leave for the hospital, she vomited. It wasn't your ordinary, everyday garden variety spaghetti-and-meatball vomit, though. She vomited shit. Her intestine was blocked, so it couldn't go down and that left one other direction—up—as in upchuck.

The smell was beyond anything I had dealt with on a call before, and in the back of that little ambulance there was nowhere to hide. How neither one of us joined her upheaval, I'll never know.

Fortunately, there was a hospital five minutes away. Unfortunately, she had to go to General Hospital in Binghamton, the emergency room farthest away, at least a twenty minute run. Unless it was a life-threatening emergency, we tried to honor the patient's request for which hospital they wanted to go to.

At the ER, we delivered her to a treatment room, and then hosed out the back of the rig. We emptied a full can of spray disinfectant in the back. It didn't touch it. We cleaned and sprayed more when we got back.

For a week afterward, every time I got in the rig, I swore I could still smell it.

About two weeks later, the next time someone started to heave in the back, the memories came flooding back, filling me with dread. When ordinary vomit came out, relief set in and I smiled. *Who knew puke could smell good.*

• • •

People do some bizarre things, and if questioned, most fire-fighters have had one or more interesting experiences with such people. During my EMS days, disco was in and dance clubs were popular gathering places, watering holes, and pick-up joints. The dress was important: silk shirts with wild patterns unbuttoned half way down the chest, tight pants, and smooth-soled dress shoes, all the better to show off your Travolta-like moves.

One local stud apparently felt inadequate regarding his "equipment," so before venturing out for an evening on the dance floor, he taped a cucumber to the inside of his thigh. The bulge was noticeable through his tight pants and apparently he believed this would work like catnip on the ladies.

The only problem was he apparently imbibed a bit too much, and didn't end up in the type of bed he was hoping for. He showed up on a gurney in the local emergency room following an ambulance ride from the club. The only ladies he impressed, with his stupidity, were the ER nurses that got to remove his artificial equipment when they stripped him down. It was a good thing he was unconscious; the laughter was uproarious.

• • •

Some of the neighborhoods we covered had their own character. The North Side was the Italian section of town. There were fantastic restaurants and the best pizza I've ever had in my life. Ambulance calls up there could be interesting, too.

We'd get a call for an elderly male patient. The main complaint could be any one of a hundred ailments, but everything else was the same. We would know immediately it was an Italian household if, as we got closer to the scene, we had difficulty finding a place to park the ambulance. Before calling us, they had to call over any nearby family members to discuss sending *Nonno* to the hospital. We'd weave our way through the crowd in the living room up to the second floor bedroom where we'd invariably find him, surrounded by still more family. After dealing with whatever medical issues were presented, we'd get ready to transport. At this point, stage two of the fun began. There was room for only one of the relatives to ride in the ambulance up front with the driver. The debate would begin. We already knew the answer and could pick the winner, but there was no sense spoiling the fun. Eventually they would decide and the result was always the same: the most hysterical daughter was selected to go along with us.

The last part was inevitable as well. As we wheeled the old guy toward the door, the four-foot-nothing, tough as nails *Nonna* would wag her finger in our faces and begin a lecture, naturally in Italian. We'd nod a lot and smile and agree with everything she said, as we got him out the door and into the rig.

Happened to me at least fifty times…

• CHAPTER SIX •

Characters

The emergency services are full of talented and unusual people and some of those people I've worked with are unforgettable. Hubie was my first ambulance captain. He was a gregarious, happy guy with a ready smile and a unique high-pitched laugh heard regularly around the squad room. He was a perfect manager for us young guys; he knew exactly how much of the reins to give us before yanking them back hard. His normal smile would instantly disappear, replaced with a cold, hard stare and raised eyebrows. You didn't quickly forget the lecture that would follow. I learned a lot about the care and feeding of young pups by watching Hubie. He never held a grudge and after whatever stupidity we had committed was over, it was over for good.

Hubie was truly a great medic; his only weakness was vomit. Not that any of us liked it, but he truly hated it. If a patient was puking, he could handle it and work through it, but you'd hear Hubie gagging right along with them.

• • •

Roddie was a charter member of the ambulance squad; he was a tall dark-haired guy whose bow-legged walk reminded you of Joe Friday. His years of service along with his personality made it impossible not to like him and bought him some forbearance for his occasional transgressions. Actually, his rule-breaking was a fairly common occurrence; the forbearance was only necessary on the occasions he got caught.

The first modular ambulance we bought back in the late 1970s, with the van front and large box behind, had what was known as a tachograph, which recorded the time and speed at which the vehicle was being driven. The captain considered it a good safety device to monitor driver behavior, which worked if the driver in question actually gave a damn what the captain thought.

One day, shortly after the new rig was put into service, the captain pulled the graph out and was surprised to see a reading of just over one hundred miles per hour. He checked the date and time this speed occurred against the shift board to determine who the driver was. To no one's surprise, except maybe his, it was Roddie.

The captain approached Roddie to question him about the speed chart. With his ever-present smile, Roddie calmly professed having absolutely no knowledge of how that speed got on the graph. They went back and forth a few times on the subject, the captain receiving no satisfaction. He knew damn well that Roddie was lying, but his options were limited. He threatened him with suspension, but Roddie knew it was a bluff. It wouldn't go over well if the captain had to suspend the lieutenant in charge of maintenance. Ultimately he gave up, warning Roddie not to "pull that shit again."

I'm not sure if the captain ever found out what the rest of us already knew. Roddie and his crew had been drag racing state troopers on the nearby George F. Johnson Highway in Endwell.

On the Saturday night eleven to seven shift, our two crews, filled with young guys, were well known for pushing the envelope as far as behavior went. One night, the two twin modular ambulances were drag racing down Main Street in Endicott, red lights flashing away. Andy was a bit heavier on the vertical pedal; that, or a bit more crazy. In either case, we won, and pulled ahead, Roddie moving over onto our rear bumper. For some reason I decided a bit more fun was in order. I grabbed the dog mace we carried on the visor and crawled to the back of the rig behind the stretcher. Sliding open one of the small windows in the back doors and with their front bumper only inches from our rear one, I sprayed their windshield with the mace, simply trying to get their attention. This resulted, however, in unintended consequences.

The reaction was instantaneous as the air conditioning unit on their rig sucked the mace in. The side windows came down and I could see them coughing and hacking and rubbing tear-filled eyes. Luckily they didn't get much of a dose. It was much more fun to watch than to be on the receiving end. Roddie has never forgotten who maced him.

Once in a while on a nice warm summer night we'd do a cookout. One of the guys had an uncle who was a butcher and would get some nice steaks. Another guy would get a dozen ears of corn. I'd get Mom to bake a chocolate cake. Naturally, Roddie was a bit more imaginative with his contribution. The Captain, as he was also known, short for Captain Carvel (he worked at an ice cream place), had keys to almost everything at the old Hooper School, then used to house various community agencies and where the Endwell rig was based. This included the kitchen and storage areas for the local Meals on Wheels group. When we were having a cookout, Roddie would acquire a tray of macaroni salad or potato salad from Meals on Wheels. We never heard that it was missed so I guess no one ever went hungry because of it.

The crews from the two rigs would meet at one of the squad rooms where we'd cook and eat to our hearts' content. It was a lot of fun. Knowing how Roddie obtained his part of the meal, I always found it ironic that he ended up as a security manager for a large defense contractor.

• • •

Big Hank was a career firefighter in Endicott, and he and his wife had been Mom and Dad's landlord when they first got married and moved to town. By the time I got to know Big Hank, it was toward the twilight of his career. He was not a small man. His ink black hair was combed straight back, and he had hams for hands; he was one of those guys who didn't remotely know his own strength. To say he handled fires the old-fashioned way would be putting it mildly.

I can't say I saw him on a lot of calls, but on the ones we did share, I never saw him in complete bunker gear. Actually, I never saw him wear more than his helmet. I'm not sure he even knew how to use an air pack.

One of the first fires I saw him on was at a big, old three story house next to a church on Nanticoke Avenue in Endicott. I was there standing by with the ambulance. Shortly after it appeared that the fire was pretty well knocked down, Hank walked out the front door, wearing just his helmet and station uniform grays, face as black as a coal miner's in contrast to the white teeth of his grin. He had on a pair of ancient untied uniform shoes. They were purposely kept untied to allow them to be kicked off quickly so boots could be put on. I don't know why he bothered, because I don't know if he ever wore his boots.

Dad always told of a snowy, frigid New Year's Day fire in the early 1960s. He was in the back of the building and watched Hank, operating by himself; advance a two and a half inch hose line up a ladder onto a porch, kick open the back door, and take it on into the building. Ordinarily, this would be a significant task for a crew of three.

Each time I saw him, his greeting was always the same. The huge paw of a hand would smash down onto my shoulder, and with a booming voice, I'd hear, "Little Ryman, how the hell are you?"

During a severe winter storm, they were calling in off-duty guys to man reserve apparatus. Typically, the chief was trying to get four or five guys on a piece. The officer at the North Side Station was checking in over the radio with the chief on the progress they were making.

"Who do you have on Engine 3?" the chief asked.

"Big Hank and a driver."

"Ok, Engine 3 is manned," the chief decided.

On a residential fire, again on the north side, they were working inside. A younger officer (they were all younger than Hank by this time) told him they needed to get water on the fire in the next room. Hank had the line in his hand, and took the nozzle and smashed it through the wall, then opened it on the fire. The officer shook his head.

"Hank, I meant we should take it through that doorway," he said, nodding toward the opening three feet away. Hank just grinned.

A big fire on Watson Boulevard demonstrated his amazing strength yet again. We were packing up once the fire was out and I saw Hank help get some empty air cylinders loaded. These were the days when heavy steel cylinders weighing a good twenty pounds a piece were still widely in use. I watched him pick up two, one in each hand; no big deal ordinarily, except Hank didn't pick them up by the handles. He picked them up from the bottom, palming the rounded end of each bottle.

One day, one of the rookies started mouthing off, giving his opinion on how the department should get into the EMS business. Hank didn't take kindly to change, and he sure as hell didn't want anything to do with "medical shit." He listened for a bit, becoming angrier and angrier.

Finally, having had enough, he picked up the rookie with one hand, by his throat, and dangled him six inches off the floor. He explained to the youngster in no uncertain terms, many of which began with the letter "f," that he hadn't been on the job long enough to have an opinion, and that he would be the first to let the rookie know when that situation had changed. In the meantime, he advised him to maintain his mouth in the closed position.

Hank encouraged the rookie to shake his head vigorously in agreement. At which point he placed the youngster again on terra firma and allowed him to resume breathing.

• • •

John B., the chief of a small local department in another area sat at the bar at the local pub relaxing after a long day's work. He was a well-known storyteller and was regaling the barmaid with some of his tales while enjoying a few cold beverages.

The telephone at the tavern rang, and the woman behind the bar stepped away to answer it. She listened momentarily and then handed the receiver to John B. It was one of his assistant chiefs.

"Stand up," the assistant chief said.

Confused, John B. answered, "What? How the hell did you know I was here?"

"Everyone in the fucking county knows where you are. You're sitting on your radio mic. Stand up!"

Luckily the stories and jokes weren't too risqué. After that, John B. didn't take his radio with him anymore when stopping for drinks.

• CHAPTER SEVEN •

First Time Inside

When I turned eighteen, I was finally able to join the fire department. I trained, studied, asked questions, and waited interminably for that first working fire—my first time inside a burning building. Months had gone by with a few close calls but never an actual working fire. Up to this point, I had been on one house fire, but hadn't made it inside.

Then it happened. I was sitting at the kitchen table sipping coffee, still in my bunker pants, when the Plectron went off. Dad and I had just returned home from an early morning call that hadn't been anything serious. As we headed for the garage, the dispatcher announced a house fire on Boswell Hill Road, a residential area in West Corners. Maybe my luck would change.

While in the car we had to contend with school bus and morning traffic. With the siren yelping and the radio cranked to full volume, we made the turn onto Boswell Hill Road and could see the house a block up the road on the right-hand side. The entire front was engulfed in fire. The porch was outlined with rolling bright orange flames spreading up the front of the shingle-covered house and towering twenty feet over the roof line. Black smoke boiled into the sky above. Dad parked the car and we finished putting on our gear as the first engine arrived. Dad had command and I stood awestruck, watching the column of fire. My breathing rapid from excitement, I got my first taste of the acrid bitterness of smoke residue that dropped into the yard surrounding the house. The excited shouts of neighbors and bystanders receded into the back of my mind as I concentrated on the crackling monster in front of me, the radiant heat on my face from forty feet away. It was hotter than opening an oven door.

The engine came up the hill, stopping just short of the house, leaving the front open for the truck. Chick, one of the two captains in the department, climbed out of the officer's seat, and he and I pulled the two and a half inch pre-connect hose line. We set up on an angle by the corner of the house so as not to hit the fire straight on and push it through the rest of the place. That was wishful thinking on our part. Older two and a half story balloon-constructed houses have lots of places for fire to spread and run.

The line went hard and I braced myself, readying the nozzle. An inch and a half hose line was also being stretched past us to a rear door. The two lines on the ground were each different lengths with different flows and friction losses. This meant that the pump operator had to remember to gate down the line or partly close the discharge valve requiring the lower pressure. In this case, that was our line and, unfortunately, the pump operator forgot.

I opened the nozzle. Chick pushed forward to take up the back pressure. Just as the heavy stream reached the burning house, both of our feet left the ground. Up in the air we went. Then, we fell back down on the ground. Hard. The wind was knocked out of me. My left hand pushed forward on the bail to shut down the line before it took us for another ride. Chick immediately knew what had happened and yelled, "Gate it down." The pump operator made corrections and we reset ourselves.

"Go ahead, hit it," Chick ordered, and I opened the nozzle again. Hitting the front of the house with two hundred fifty gallons per minute quickly extinguished the exterior flames and shortly thereafter turned the front of the building black. Smoke, still black and viscous, continued to pour out from the windows, evidence that plenty of fire was left inside.

We dropped the line and I ran over to the squad and grabbed an air pack out of the side compartment. We were to take a line in through the rear door and push the fire back out the front. Dad was busy managing the overall fire attack. He didn't supervise the placement of every line and ladder. That was the responsibility of the lieutenants and captains.

The other firefighter about to go in with me was also new. As rookies, we had no business being inside together, but I had waited too long for this moment to complain. Taking the nozzle in my hands, we crawled inside. It was black to the floor with smoke, but there was no visible fire in the first room we entered. We worked our way in by feel, stopping so I could work on what felt like a door. Still zero visibility; the only sounds were the raspy rapid inhalation and exhalation of air through our mask face pieces. Prying for a minute or so, I got the door open and realized it was a kitchen cabinet. We pushed further in and through the doorway to the living room. This room was fully engulfed in flames. I opened the nozzle and started hitting it, whipping it in circles as I'd been taught. *Damn, this is fun.*

The next thing I knew, the fire rolled over us from our left. Someone had opened up a two and a half hand line from the side door, but I didn't realize it at the time. We fell assholes over elbows back into the kitchen. If I'd had more experience, hell any, I would've known what was happening and dropped the stream down to pop the other crew and let them know that we were there.

Back in the kitchen, as we tried to figure out what the hell happened, a captain crawled in behind us. "Who do I have?" he called out through his face piece.

"Ryman," I answered, and the other guy called out his name.

"What the fuck are you two doing in here together?" the captain roared back, a rhetorical question at this point. The other line had shut down and we regrouped and advanced again into the living room, killing the visible fire; the remains of the room were a charred, blackened, and steaming mess. Our low air alarms began to ring, and it was time to leave.

Two more air cylinders, and hours of ass busting overhaul and clean up later, we finally got home. The house that burned was pretty much totaled, but I was too excited to be tired. I'd had a good taste of a working fire, and wanted more.

• • •

The fire drought had ended and my next fire came the following week. It was even better. The old squad truck was fun to ride in; it was a rear walk-in with a dark, narrow aisle up the center; an early model of its type. The squad crew was designated to make the initial attack or rescue, depending upon the situation. The engine crew would stretch the line to whichever door and the squad crew would stroll up and take it in. It became my favorite position to ride.

We were on our way to a mid-morning, weekday house fire on River Road. Hal, my partner that morning, and I dismounted the rig as we lurched to a stop behind the engine. Smoke poured out from every opening in the two story house. It was black, viscous, and pushing hard—indicative of a lot of fire. The engine had laid a supply line from the nearest hydrant, and the crew pulled a two and a half hose line for us to the garage door. This place was ripping, and there were only six of us.

I knelt down, tightened the straps on my face piece, checked my gloves, and waited for water. The line grew hard, and I picked up the nozzle. Hal tucked in behind me. We advanced into the garage, crouching over, visibility near zero, working our way in until we could see the glow in the corner. I opened the nozzle, moving the heavy tip as much as possible as Hal pressed in tight behind me, absorbing the back pressure. With some snapping and sizzling, the glow died, and we prepared to advance further into the blackness.

From above our heads, we heard a noise—gunshots. Lots of them. "What the fuck is that?" I shouted. "Let's get the fuck out of here!" We backed the line out of the garage. Later in my career, I learned that ammunition in cardboard boxes was not a big problem, but this was my second time inside, and all I knew was that World War III was going on over my head.

Once we were outside, we had a number of choices other than a Normandy-like invasion of the garage. The front door was closest to us so we popped the door and crawled into the living room, the fire greeting our arrival. I opened the nozzle again, swinging it in circles at the ceiling, controlling the flames in the room. Having killed the living room, we worked our way toward the kitchen as our bells began sounding, signaling our air supply was nearly depleted.

We again backed the line out, but there was no one to relieve us, not unusual for a weekday fire. We dropped the line to go change our air bottles. The fire wasn't waiting around for us, and by the time we got the packs off to change our cylinders, it was obvious we weren't going to catch this one. The fire was probably on both floors when we pulled up, but I didn't know that then.

Help began to arrive shortly thereafter. Our neighboring company arrived with their engine, and Rodney, the chief at the time, had called IBM, the area's largest employer, to release our members. This was a bit of an unusual arrangement and was seldom used. Along with most of the other local departments, we had an arrangement with IBM to release our members from work in the event of a major fire. The IBM security department maintained lists of various departments' firefighters and their work extensions. When notified by communications that a department requested their members, security would contact them and tell them the location of the fire.

That was the reason why, after my second air cylinder, I came out and saw my father in the front yard getting some crews organized. There were enough people around now that we didn't have to change our own bottles. I took a few breaths, and bent over a bit while somebody removed the empty air bottle and replaced it with a full one. I was pretty tired at this point but was only eighteen-years-old, so my body was resilient even if I was young and stupid.

With the third cylinder on my back (these were steel or aluminum, not the feather-weight composite ones we have today) I attached myself to a crew advancing a line in through the garage. The nozzle man and backup were on the floor above us, and I was feeding them line from the bottom of the stairs. I didn't remember to put my collar back up before I returned inside, and bending over to advance the line upstairs exposed a bit of my neck. Scalding water coming off the fire on the floor above me ran right through the floor boards and down across my exposed skin. The pain wasn't enough to force me right out of the building, but I was pretty sure I had been burned.

When that cylinder emptied, I went out and dropped my pack by the squad. I looked around and found Dad, still directing operations in the front yard. I walked up to him and took off my helmet.

"Take a look at my neck. I think I got burned," I told him. He pulled my collar back and looked at it.

"Yep, you got burned," he said. "You'll live. Get back in there."

That was all the sympathy I got. Anybody else would have gotten the rest of the call off and maybe a check by the ambulance crew.

Young, foolish, and having too much fun to protest, I looked around for some other way to work the fire. The house was pretty well trashed, with little or nothing to save at this point. It had been too far advanced on arrival for us to stop it with the limited manpower we had.

Dad had me take a line around the right side of the structure and, for whatever reason, he decided to tag along. There was a room off, fire blowing out the window, and nobody inside opposite it, so he told me to hit it from there. I opened the nozzle and knocked the fire back, moving up to the window. Shooting a straight stream at the ceiling, I rested the line on the sill, letting the droplets coming off the ceiling do the work.

Dad was behind me. "C'mon, move that goddamn thing around," he bitched.

"I'm too damn tired to move it. You want it moved, you do it," I whined over my shoulder. So he did. He took the nozzle and got it going in a circular motion at the ceiling. I think he was secretly having a good time as chiefs seldom got to play, particularly with the nozzle.

Hours later, I rested on the couch in the family room at home. I had been bitching for weeks about how little fire I was getting. The same thing Mike does now. Dad walked into the room and stood over me.

"Seen enough fire now?" he asked.

"For a day or so," I answered. He smiled and walked away.

• • •

A few weeks later, we were hit for a structure fire in the middle of the night involving a vehicle repair garage attached to an older two story house. As Dad and I pulled in, we could see the flames through the roof of an addition in the rear of the garage, lighting up the scene like daylight.

Dad picked up the radio mic and said, "Car 38 on scene with a working fire."

Dennis, the fire marshal, pulled in right behind us. Dad started to size up the building to set up the initial attack while Dennis and I headed for the house to make sure everyone was out.

Entering the front room, we found the owner—a classic gray-haired little old lady—chasing a dog around the furniture. How the dog, or the lady, could move anywhere in that house amazed me. She was a pack rat. There were piles of magazines and newspapers everywhere. The stairway to the second floor had about a six inch path up the center. There were a few narrow trails through the remainder of the living room. We were in deep shit if the fire made it into the house.

"Ma'am, you've got to leave," I told her.

She continued chasing the dog, ignoring me. Dennis cracked the door from the main room into the garage. Black viscous smoke immediately pushed into the room. He slammed the door and looked at me, his wide eyes telling me to hurry up.

"C'mon lady, pick up the dog and let's get out of here."

She continued her pursuit. The dog ran in my direction and I was able to scoop it up and hand it to her.

"Let's go," I told her, motioning her toward the front door.

"I have to get my purse," she now changed the object of her search.

"Forget the purse, we are leaving now," I told her, my voice just under a yell, starting to get worried.

Finally she acquiesced and we were able to get her out of the house.

Back out front, I could see the fire spreading further into the garage. The engine was coming down the road, siren screaming, red lights bouncing through the darkness. The big American Lafrance pumper stopped just past the building, leaving room in front for the truck; three inch hose peeled off the hose bed and back to the hydrant.

I ran over to the engine to grab an air pack. Dad gave the officer on the engine a few quick instructions, and pulled the two and a half inch pre-connect himself. Another arriving crew was directed to pull an inch and a half and vent the garage roof.

While that second crew, to which I was assigned, got ready to open the roof—putting on air packs, getting the K12 circular saw, stretching the protective line, throwing ladders—Dad flaked out the two and a half inch pre-connect on the side of the building. He twirled the forefinger of his gloved hand over his head, signaling the pump operator to charge the line.

I knew what he wanted to do. Where the line was positioned, its crew could push the exterior fire back and possibly get some water down on the now fully involved rear addition. Dad hoped that might slow some of the interior fire spread in the garage.

The only problem I saw was that the two and a half didn't have a crew; instead, Dad stood there by himself. Typically, you would have two or possibly three guys on such a line; more, if you actually wanted to advance while operating it.

None of that stopped my father. Nozzle in hand, he braced himself, and opened it up. What followed was a straight stream, two hundred fifty gallons a minute flowing toward the rear of the building, across the roof, darkening down the fire showing in the rear addition.

We were up on the main roof by this time, and Riley from the squad cut the hole with the screaming circular saw while I covered the operation with the hose line. I looked down at the side of the building as two guys relieved Dad on the line so he could get back to running the fire. They opened the nozzle, and got knocked on their asses. I started laughing through my face piece; not at their misfortune, but at the irony of watching one little guy with lots of determination hold the line by himself, and then watching two big guys fly through the air.

The roof now open, smoke and heat released through the hole, we climbed down as the rest of the attack began. With the garage vented through the hole we cut in the roof, the overhead door was opened and wedged up, and two crews, each with an inch and a half line, pushed in. The fire was out about three minutes later.

The pack rat lady's house was fine. She could go back to her newspaper collection and chasing her dog. If the fire had gotten in there, we would've been working on it for days. The most enjoyable part, though, had been watching Dad operate the big line by himself. Pretty damn good for a guy three inches over five feet.

● ● ●

Day crew training was always fun. There were fewer people and no chiefs around, so we could push the envelope at times.

During one training session, we were discussing the possibility of operating a line from the top of a 50 Bangor ladder in what was known as an auditorium raise. With that technique, the ladder is vertical, not leaning against a building, and held in that position by poles, ropes, and lots of muscle. We brainstormed ideas on when operating a line in that position might be remotely useful. Auditorium raises are typically used as a confidence-building drill in training rather than an actual firefighting technique. There are few practical uses for one otherwise. The standard drill itself consists of climbing up one side, throwing your legs over the top one at a time, then climbing back down the other side.

While we may have been crazy, we weren't stupid; we decided to test the theory incrementally. Instead of using the bangor, we decided to use a thirty-five foot ladder instead. We attached two ropes to the tip of the ladder tying the knot in the center of each rope, giving us the needed four ends. We carefully completed the auditorium raise, one firefighter on each beam, and one on each rope end, steadying the now perfectly vertical ladder.

I eyed the slightly swaying top, threw the inch and a half nozzle over my shoulder, and began to climb, the hose trailing behind me. It didn't take long to reach the top. The plan was to charge the line and see if I could operate it from the top. I pulled the nozzle off my shoulder in preparation for the line being charged, and held on for dear life.

Chick, the captain who was supervising the drill yelled up to me, "Put a leg lock on." Chick was a tall guy with frizzy hair, so much of it that his helmet barely fit.

I pretended not to hear him.

He yelled again, "Put a leg lock on." This meant putting my leg through the ladder rung and back out and locking my foot on the beam to secure myself to the ladder. This procedure, while normally a good safety practice, had in this case what I believed to be a few drawbacks. I knew I needed to explain these potential issues, at least as I saw them, quickly and succinctly.

"Chick," I yelled down, "with all due respect, fuck you. I wanna be able to jump if this thing goes over."

The highly trained professional that he was, Chick carefully considered my technical feedback on the scenario. He called back up to me with his decision, "Okay, but you better not fall, or your mother will kill me!"

• • •

While attending the local college, I worked part time teaching first aid and CPR classes to commercial clients for the university. I was working on my Associate's degree, trying to decide what to do for the next forty years. I knew finishing up my Bachelor's was the next step, but after that I wasn't sure; maybe a job as a career firefighter.

If I went that route, I knew I wanted to ultimately become a Chief. As I was to learn, my education would be of little help with the civil service at the lower ranks, and the politics at the senior levels.

On the way home from teaching class one afternoon, I heard a brush fire come in along Nanticoke Drive. I actually enjoyed brush fires back then, the same way Mike does now. I went to the station and caught the last piece to the scene. The fire wasn't huge, but it was a pain in the ass to work, as it was on a steep hillside. We actually had to rig a few rope lines to get up to it and work laterally along the head of the fire.

Our personal protective equipment requirements for brush fires in those days were boots, helmets, and gloves. I was wearing a nice light blue dress shirt. By the time we had the fire out, it wasn't entirely blue anymore, or nice.

We returned to the station and got the tools cleaned and the booster tank filled, and everything back on the apparatus. I was ready to go home.

"You ready to go, Dad?" I asked.

"Uh, how about a beer?" he responded.

"No thanks," I answered.

"No, c'mon and have one," he insisted.

"I don't want a beer. I want to go home," I explained.

By this time I was starting to get the impression he was avoiding home. I gave him a look and said, "Okay, what's the deal?"

"Your Mother's going out in a little while and this way she won't see that shirt of yours until tomorrow," he admitted. He'd be gone to work then and he knew he would get more grief about it than I would, so we stayed and had a beer together.

His plan backfired, though. When we got home Mom hadn't left yet, and yes, he got an earful.

•   •   •

It was around suppertime when the Plectron went off for a reported kitchen fire on the lower section of Jeannette Road, the same street we lived on. Dad and I jumped in his car, turned on the light bar and siren, and started down the hill.

The house was on my side of the car, so I saw it first, and keyed the microphone: "Car 30 to communications, on scene, smoke showing." Dad still hadn't seen it. "Where?" he asked.

"The house over there with the smoke coming out the front door," I responded, somewhat of a smart ass. He pulled the car into the driveway of the house across the street from the burning home. We each started putting on our bunker gear. Dad continued his size-up, picking the hydrant he was going to have the engine lay from, and considering the apparatus positioning, along with anticipating the fire's location, its size, and growth.

I looked at the smoke conditions. It was pretty snotty and black as it rolled out the front door, but was banked nicely, about four feet thick at the ceiling level. That left the floor level relatively clear. I grabbed the five pound dry chemical fire extinguisher that resided in the trunk.

"I'm going to give it a shot and see if I can get it," I told Dad.

Hearing no response, and taking that as permission, I entered through the front door, duck-walking to stay low but quick, through the living room and into the kitchen. When I entered the kitchen, I could see that the entire surface of the stove was on fire, along with the two cupboards directly over it and the cupboards on either side of these. Flames rolled across the ceiling toward me.

I didn't hold out much hope that I could put it out, but I knew I could sure as hell slow it down a bit. I pulled the pin on the extinguisher and lifted the hose up toward the edge of the fire on the ceiling, and squeezed the handles together. The dry chemical began to discharge, and as soon as I hit the rolling fire, forcing it back, I immediately shifted to the stove itself, the seat of the fire.

My efforts were having a good effect on the fire, but they also caused the thermal balance to shift, so that all the smoke and heat dumped right on top of me. It was black now, right to the floor, which is where I put my nose in an attempt to get some clean air while I continued to empty the extinguisher. I kept the nozzle moving in the direction of a fire I could no longer see.

Just as I ran out of powder, I felt a hand on my coat. Dad had crawled in behind me. We were only an arms length apart, but we couldn't see each other.

"What do you have?" he asked.

"The stove and cupboards were going good," I told him coughing through the smoke. "We need a line in here quick 'cause it was in the hood real heavy." I was concerned about fire extension through the ventilation hood over the stove in to the attic.

"Okay," he said. "Let's get the hell out of here."

We crawled into the living room, and out the front door. A couple of breaths of fresh air once back outside, and I felt pretty good.

Dad had the engine, which was now arriving, stretch an inch and a half inside to finish off the fire and check for extension. A three inch supply line from the hydrant to the engine snaked down the street.

After about five minutes, the engine crew came out onto the front lawn, and began removing their air packs. The captain reported to Dad, "Fires out, Dick, no extension either."

"How much water did you use?" Dad asked.

"Not a drop," he replied. "It was out when we got in there." The extinguisher had gotten the whole thing.

• • •

We were dispatched for an investigation of a fire reportedly out—a "fire out" call. Most of the time they actually were out, but sometimes they weren't, and we couldn't take a chance. A full structure assignment would be coming unless the first Chief on scene directed otherwise.

The house was around the corner from ours, so it didn't take much more than thirty seconds for Dad and I to arrive. One thing was instantly clear—the fire definitely wasn't out.

It was a two story home with an attached two car garage. Smoke oozed from beneath the closed garage doors. There was an excited teenage boy in the driveway waiting for us.

"Car 30 on scene, smoke showing," Dad called over the radio.

We put our gear on and started to investigate. The garage doors weren't hot, a good sign. I lay on the ground and Dad raised the door a couple of inches. There was plenty of smoke, but no visible fire.

"Put it up all the way," I told him. He did and propped it open with a shovel lying nearby. Garage doors have been known to close on firefighters, trapping them in the building, if the springs weaken due to heat.

It got weirder. The smoke appeared to be coming out of the floor. We knew there were no basements beneath the garages in these houses. We went in further to see what was going on.

Holy shit. There was a grease pit in the floor. I'd seen one in my Grandfather's old repair garage years before, but never any place else, much less a house. It was about ten feet long, and about five feet deep, about a shoulder width across. There was a ladder at the far end. The smoke lifted nicely by this point with the garage door open. It seemed to be coming from the end of the pit opposite the ladder.

I looked at Dad and he nodded, so I started down the ladder. Reaching the bottom, I worked toward the end the smoke was emanating from. Duck-walking across the floor of the pit, I reached a metal box, about three feet square, that had smoke billowing from it. I didn't know what the hell it was. The first engine arrived by this point, so I asked Dad to send a line down. He lowered it to me, and I soaked the box, getting water into as many openings as I could see in the partial darkness of the pit.

The smoke died down some, and with the help of a couple other guys and a roof ladder which they lowered into the pit, we lifted the box up so it could be taken outside. By the time we got out of the pit, the box had been removed from the garage and a couple guys had opened it up to complete the extinguishment. Old rags, scraps of wood, and similar items were being pulled out onto the driveway, some burned, some still smoldering.

Dad talked to the kid and the use of the box became clearer. It was an old dishwasher. They loaded it up with the wood and rags and placed it at the far end of the pit which had been turned into the family shooting range. They placed targets on the outside and shot .22's into it as the backstop to absorb the rounds. Unfortunately, enough heat had been generated by the rounds to ignite the combustible material inside.

The original owner of the house had been a car enthusiast who had the grease pit constructed when the house was built. These folks had purchased the home subsequently and found a new and more exciting use, at least that day, for the grease pit. It was back to the drawing board, though, for the intrepid marksmen, as they now needed a new backstop.

• CHAPTER EIGHT •

Working for Dad

About a year later, I made lieutenant at the tender age of nineteen. It was still the early 1980s and I didn't know at the time I had too little experience for that role. I had to learn to manage individual operations on the fire ground and command small crews, putting aside my black helmet for a yellow one. Shortly after putting on the yellow helmet, a late night barn fire was dispatched.

Dad dropped me off at Station 2 on his way to the scene. I rode the officer's seat of the snorkel, which was an articulating platform using two large arms. We positioned on the main road, just down from the house, so we could protect it if we lost the building. There was heavy smoke showing from the loft level of the old red barn. Engine 33 took the driveway between the house and the barn. Unfortunately they didn't lay a supply line as they drove up the driveway, which would create a bit more work.

I grabbed a pack off the engine. Dad had a couple of guys start to hand jack a supply line back down the driveway—about fifty feet of line over each one of their shoulders—while an inch and a half was stretched toward the barn. Engine 37 came in at the bottom of the driveway. They would supply Engine 33 from their one thousand gallon tank as well as from the tankers Dad had coming due to the lack of hydrants in this area.

I donned my face piece and followed my nozzle man Dennis into the building. It was my first fire as a lieutenant in charge of an interior attack. In theory, I was in charge, but at least smart enough to know that Dennis had more time on that nozzle than I had in the department. I wasn't about to give him any advice.

We knocked down a little drop down fire on the ground level and then headed for the stairs to the loft. At the top of the stairs we made the turn onto the heavy plank floor, and found the smoke banked right down. We started in, tossing hay bales out of the way as we went. A few feet in, we could see the fire and Dennis opened up on it. We pushed across the floor until it looked like we had it all.

I crawled back to the top of the stairs to pull a bit more line up. I noticed a white helmet watching me from the top of the stairs.

"What are you doing up here?" I asked, my voice muffled through the face piece.

"Just trying to find out what's going on. Not getting any good information out there," Dad responded.

I made a smart ass comment about his lack of an air pack. He tapped the front of my helmet.

"What does that say?"

"Lieutenant," I answered.

He tapped the front of his helmet. "What does this say?"

"Chief," I answered again.

"Go back and do your damn job," he ordered. There wasn't much to add, so I pulled up another five feet of hose and went back to Dennis. I never did understand how he could breathe smoke like that and not cough. I didn't buy his desire for more information; he was checking up on how I was doing, but he would never admit it.

• • •

I was on my way home from teaching a CPR class, while Dad was out teaching one of his state fire courses. Driving up Route 17C toward Station 2, I heard the call for the kitchen fire come in over the scanner in the car, and recognized the Glendale Drive address as Don and Nancy's house, very old friends of our family. Knowing the people involved in an emergency gives any firefighter an extra shot of adrenaline, and I got one, as I sped to the station.

In the station parking lot, I donned my gear and climbed into the front seat of the truck. The eighty foot snorkel was a blast to ride. Dad enjoyed pointing out it was an "articulating platform" to those who used the generic name.

Mason, the driver, threw me his boots as he climbed up into the cab. He was an old, bald, curmudgeonly guy who now only drove; his helmet the only protective gear he typically wore, more in deference to the cold in the old-style no roof open cab on the truck than any rules. As far as Mason was concerned, that was all an officer on the piece was good for: holding his boots, particularly when that officer is young enough to be his grandson.

As we pulled out of the bay, I could see the engine crewing as well. It looked like we would have plenty of manpower.

Mason made the right onto Route 17C, and then the left onto Glendale Drive, the long white booms of the aerial shining in the moonlight, red and white lights bouncing off them from the two beacons over the windshield. My foot pressed hard on the pedal on the floor, filling the night air with the siren's scream.

Brian, the assistant chief, arrived on scene, and slowed everyone down. The fire was out. We had a bit of smoke that would need to be exhausted from the house, which was good news. Mason pulled in nose-to-nose with the engine out of Station 1, which had come from the other direction.

I went into the house, glad that Don and his daughter, Kelsey, were okay, and that the damage was minimal. It didn't take long to get things cleaned up and the smoke cleared out.

It wasn't until I got home that I found out about the unusual route the emergency call for the fire had taken. In the early 1980s before 911 was widespread, many departments, ours included, had individual seven digit emergency numbers. The system still worked well, if people used it. In this case, it worked a bit differently.

Mom was home, watching television, when the phone rang. She answered to find Kelsey on the other end.

"Hi Jeanne," Kelsey said. "Is Rich there?"

"No honey, he's not," Mom answered, a bit confused as to why Kelsey would want to talk to Dad.

"Is Gary there?" Kelsey followed up; always nice to be the second choice. Now Mom was really confused.

"No, he's not here either. What's wrong, honey?" She asked, knowing something unusual was going on.

"Well, the kitchen's on fire," Kelsey answered.

It was time for Mom to ride to the rescue.

"Okay Kelsey, get out of the house. I'll take care of it," Mom told her.

She disconnected and quickly dialed the emergency number on the sticker attached to the phone. The dispatcher quickly picked up.

"Fire department, what is your emergency?"

"Send West Corners to 322 Glendale Drive for a kitchen fire," Mom instructed.

"Do you see fire, ma'am?" the dispatcher properly questioned.

"No I don't see fire. I'm not at that house," Mom replied, starting to get a tad impatient.

"Well, then how do you know there is a fire?" the confused dispatcher asked.

Mom recognized the voice as Karin, one of the regular dispatchers. She had worked as a nurse years before with Karin's mother. Karin obviously knew Chief Ryman. Mom decided to put an end to the twenty questions game right then and there.

"Karin, this is Mrs. Ryman. Just send the fire department now!" she told her. "Okay Mrs. Ryman. Right away." Click went the phone line. About five seconds later, Mom heard the Plectron tones go off and the same dispatch I heard. It was one of the few times Mom thought it was good to be the chief's wife.

•　•　•

The Friday afternoon of Memorial Day weekend in 1981, the temperature was more like late July than late May. I was helping Dad with a few things around the house, when we heard Endwell get hit for a structure fire on Watson Boulevard on the scanner. Soon after, our tones went off on the Plectron for the truck to assist.

The dispatcher sounded just a bit excited. "Attention all West Corners fire personnel, Truck 36 to assist Endwell on Watson Boulevard for the apartment building fire."

Dad and I jumped in the car, flipped on the lights and siren, and headed for Station 2 where the truck was based.

We were the first two there; we opened the overhead door for the aerial and I climbed into the cab. I flipped both batteries and the master switch on, with one push of the ignition button the diesel engine roared to life. I reached over and threw the switches to activate the red lights, then got out to wait for additional manpower and to finish putting on my gear.

A minute or so later, Chick pulled in, followed by another guy. We decided not to wait any longer and the three of us crewed the piece with Dad staying behind. Chick had more experience driving the truck, so even though he outranked me, he drove and I rode the seat.

With the siren screaming, we started down Main Street and over Nanticoke to North. It was a long run, and hitting North Street just as the IBM employees were getting out of work made it interesting with the congestion.

Starting down Watson, we could see the jet black smoke looming up. That made it clear they didn't have much water on it yet. We were one of three additional trucks on the way, so that would change shortly. As we got closer, we could see three apartment buildings with fire through the roof. I pressed the siren pedal harder, my foot trying to push it through the floor board.

There were actually five attached buildings in this complex. Our job was to try to keep the fire from spreading to the other two. We turned into a driveway, taking us behind one of the exposure buildings. Chick backed the truck toward one of the three burning buildings. While the outriggers were going down to stabilize the aerial, I talked with the engine officer tasked to supply us.

"What do you need?" he asked.

"Three lines, if you can give 'em to us," I answered. He trotted off toward his piece and crew to start getting supply lines for us.

I climbed up and into the bucket, and took the pistol grip controls in my hand while looking around to make sure we were clear of obstructions. Chick climbed up to the turntable to monitor the controls there.

Squeezing the trigger activated the throttle and sent power through the hydraulic lines. I pulled up for a second or so to raise the bucket a few feet then pushed forward on the handle to raise the booms, both upper and lower, out of the bed.

The engine guys were dragging lines over and once I raised the bucket up a little, they tied into us, and charged the lines.

We pulled face masks out of a compartment in the bucket, and plugged them into the pre-plumbed air supply. We didn't need them yet, but if we had to go out over the building, they were ready.

I raised the bucket up above the roof to get a look at conditions. Fire blew through the roof in most of the building in front of us, a solid column of flames steadily working down the length of the building almost reaching its neighbor. My first objective was to knock it down near the juncture of the adjoining building, hoping we could stop it from spreading further.

I pulled the lever on the monitor nozzle, and seven hundred fifty gallons of water per minute began to flow through the big fog nozzle. Adjusted down to a straight stream, we poured it onto the fire in the corner of the building, knocking the snot out of it. With that section controlled, I moved the bucket slowly to the right and we worked over the rest of the attic level with the big gun.

That helped, but did not control the fire still raging on parts of the second floor. I dropped the bucket down slightly below one of the second floor windows and opened up again on straight stream, breaking the window in front of us with the force of the water. The powerful stream bounced off the interior ceiling, spreading water droplets and knocking down fire.

Window by window, we worked the entire building in this manner until the bulk of the fire was killed.

A pipe shaft on the corner of the building was being a pain in the ass. It was acting like a chimney flue with fire traveling its entire height. We couldn't lob our stream into it effectively. It wasn't a huge problem, as the fire in the shaft wasn't going anywhere but straight up, but it was pissing me off that I couldn't get it out.

Finally, a light bulb went on and I knew how to knock it down. We put on our masks, as we needed to be over the building for what I had in mind. Chick and I had steadily conversed over the intercom between the bucket and the turntable for most of the fire, but I didn't tell him what we were about to do. Forgiveness is easier to obtain than permission.

The bucket was equipped with sprinkler heads beneath it in case you had to go over a burning structure. It was designed to keep things cool beneath your feet. I had a different purpose in mind.

I maneuvered the bucket over the shaft so it was directly beneath us, and then pulled the lever to activate the sprinklers. I then dropped the bucket closer to the top of the hole and watched through the grated platform we stood on as the sprinkler water worked down into the pipe shaft, knocking down the fire.

Damn, this was fun. Kick-ass *exterior* firefighting. The guys on the ground were busting their asses, stretching big lines, moving them around in this heat. We were up here above it all spraying water, knocking the hell out of the fire.

The fire in the building in front of us was knocked down. I looked around for something else to do. Across the courtyard was another building being attacked by Endicott's stick. Big Hank was running the turntable on their ladder. The Chief had pulled one of the regular crew and replaced him with Hank when he heard the dispatch. He knew Hank could put that ladder up a bat's ass in a windstorm if need be. Part of the roof on their side was still intact, shielding the fire beneath from their stream. From our vantage point, however, we could see it rolling.

I took the bucket up to about seventy feet and reopened the gun directing it across the courtyard on straight stream. The guy on the top of the stick gave me a confused look. Then he saw the steam roll out from the underside of the roof where we were hitting. His face lit up and we got an "okay" thumb and forefinger gesture from him.

With most of the heavy fire in the three buildings knocked down, crews were preparing to enter to begin overhaul operations. Down below me, Jackson, a friend of mine from Endwell, was getting ready to open the door to the first floor from a small porch.

I yelled down to him, "Jackson, don't open that door." No response. "Jackson, if you're gonna open that door, stand to the side." Still no response.

He opened the door, and as I had anticipated, got hit with a chest-high wave of water. Hell, we'd been sending thousands of gallons into the second floor. I knew much of it had flowed down, since that's how gravity works.

I took the bucket back to the ground, and Chick and I traded places so he could see it from above. Now down on the turntable, with the adrenaline from the excitement wearing off, I started to feel chills. In over ninety degree weather, this meant only one thing: early heat exhaustion symptoms. I stripped off my coat and another firefighter handed me something to drink while I manned the controls. It helped.

After a while, we went inside to look around. It was a good stop. The original buildings were heavily damaged, but keeping the fire from spreading to the other two was a big win.

When we were released, we drove over to Endicott's station to fill the air cylinder. Big Hank had been on their aerial at the fire and they were already back. When we walked in I received my usual greeting.

"Little Ryman, how the hell are you," he said as his ham-sized hand slammed into my shoulder.

It's pretty bad when your body is more sore from a "hello" than an entire afternoon of fire.

• CHAPTER NINE •

A Move to Maryland

After finishing my Associate's degree in Fire Protection Technology from Broome Community College in June of 1981, it was time to move on to higher education by attending the University of Maryland to study Fire Science. In order to have a place to stay, I had arranged for a live-in slot at a fire station in Montgomery County, Maryland. What I didn't know at the time was my real education was about to occur for the most part—not on campus—but in that fire station.

Wasted college fuck. That was one of the nicer names I was called when I moved into the fire station in Montgomery County. The career guys put the new college live-ins or bunkers through a sort of mental boot camp. The first semester, I was the only "wasted load" at the station. If you can't take the verbal abuse, you certainly don't belong there.

I understood it then, and to this day, I have no problem with it. Look at it from their perspective, every year or two some strange new kid shows up. After he familiarizes himself with the equipment on the engine and truck—in my case it took about forty-five minutes—this rookie is riding a jump seat alongside them. In their shoes, I wouldn't trust a *wasted college fuck*, either.

I realized the best thing to do was to keep my mouth shut and ears open. I gained their trust by doing anything they asked around the station and doing a good job on my first few fires. Once they could see I wasn't going to get one of them killed, they began to accept me.

• • •

The twenty-two story high rise at 11700 Old Columbia Pike was a very common call. Alarm bells, odor of smoke, food on the stove, and the occasional real fire were regular occurrences there. Still new in the station, I didn't know any of this. I was in the left well of the wagon (short for hose wagon, as the engine was sometimes called).

We were the second due engine and the truck was the first due aerial. We positioned the engine and entered the building from the rear. I had the tools and the purse; the other guy had the stand pipe pack of hose. The tools consisted of a Halligan, an axe, and the K-tool for forcible entry. The purse was a bag full of fittings from which we could rebuild a stand pipe connection if the components had been stolen or damaged; not uncommon in these buildings.

We followed the engine officer into the stair well and, to my surprise, we went down, not up. Reaching the bottom, the officer felt the door for heat. When he was satisfied, he cracked it open. It looked like the lights in the hallway were off. Wrong again. It was smoke, black right to the floor.

He closed the door while we put our face pieces on and flaked out the line, connecting to the stand pipe outlet. We opened the door again and advanced the line, the officer leading us into the blackness. The nozzle man and I stopped at a point where the hallway split. The officer went one way, and the other guy went the opposite way, each looking for the fire. A minute or so went by until we heard the officer's muffled holler through his face piece for us to come in his direction. He had found the fire.

We could feel the heat start to build as we advanced the charged hose line down the hall. Reaching the officer we could see the glow at the ceiling as the fire rolled down the hallway toward us. The nozzle man opened the line at the ceiling, and I braced him, pushing forward on the line and his back. It didn't take long to knock the fire down but damn it was hot. Firefighters almost never talk about the broiling heat inside a building. After a serious, hot fire, you will likely hear the masters of understatement comment that "it was a little warm in there."

The truck crew came in behind us to pull the ceiling, and the heat actually kept them on their knees as they used their hooks. Down in this basement, there was no place for the smoke to go, so it took forever to lift.

Once we could see a bit better, it became evident what we had been up against. Somebody had dragged a mattress into the hallway, poured gasoline on it and lit it off. It was my first working fire in Maryland, and a start at showing the career guys I could be trusted.

• • •

Shortly after shift change one morning, the engine and ambulance got hit for a female down from an assault in the laundry room of a nearby townhouse complex. We could tell it was bad when they held us out on the main road until the police got on scene. I rode the engine this bright sunny morning with three career personnel.

We pulled in behind the ambulance, and seemingly a hundred police cars. There were cops running all over with guns drawn, looking for the perp. Normally we would ditch our coats and helmets on a call like this, but not that day. Nobody wanted to be mistaken for the bad guy.

We made our way to the laundry room at the rear of the row of townhouses. The door open, we could see the victim lying in a pool of blood on a cold concrete floor. Her throat had been slit from one side to the other.

We found out later that the asshole that killed her had spent the night before playing with his Buck knife cutting up beer cans and watching *The Texas Chainsaw Massacre* over and over. He then went out to play for real and found this unfortunate woman simply doing her laundry after her kids went to sleep. The poor kids discovered her the next morning.

It didn't seem like such a sunny day anymore. My first murder, but unfortunately not the last. I kept thinking about those poor little kids finding their mother like that and what their life without her would be like. Back at the station, we spent an hour or so around the kitchen table shooting the shit. It is how we deal with calls like this. Nobody spoke of feelings or emotions; it wasn't done that way. Just revisiting the call and talking through the events helped, and after a while, we got up and went about our business, in my case studying. We never knew; another call could come in the next ten minutes.

• • •

Fat Head was an interesting character. He was one of the more outspoken, opinionated firefighters I had met. He commented regularly on the state of the department. A regular refrain was that he couldn't wait until he had his twenty years in, making him eligible for retirement—which was getting close—then he would really start speaking his mind. Most of the officers were looking forward to that as much as a visit to the proctologist.

Fat Head was the truck officer on a call for a cellar pump. Cellar pumps are not a lot of fun, but they can be an important public service. It was a cold winter day when the ladder truck pulled up in front of a well-kept residence. (The portable pump was kept on the ladder truck.) Fat Head walked into the house and was met by a frantic woman who was pointing down the cellar stairs.

Based on her behavior, he expected a cellar inundated with multiple feet of water from a frozen and broken pipe. He was a bit confused when he reached the bottom of the stairs, feet still dry, and could see no water at all. He started walking through the cellar and eventually found a small puddle of water a couple of feet across.

He returned to the first floor to find the still excited homeowner. "Lady, the only water down there is a little puddle."

"Yes, that's it," she said.

"Don't you have a mop, lady?"

"No," she responded. "That's why I called you." Well, Mount Vesuvius erupted.

"Lady, you see that truck out there? It ain't got no roof. It's freezing out there and you call us for this?" He continued, leaving no time for a response. "Don't you ever call us again unless your whole house is on fire. I don't mean just the upstairs. If it's just up there, you wait till it gets down here!" He stormed out of the house.

I never did find out if she complained. It wouldn't have mattered to Fat Head; he almost had his twenty in.

· · ·

The ambulance, or "amblance" as they were known in Maryland, was sent to neighboring Station 15 for a walk-in. Their own "amblance" was out on another run at the time. Somehow, Fat Head had ended up on the call. How, I don't know, because he hated running on the ambulance, and technicians like him ranking above firefighters, were usually exempt from that duty.

On arrival, they found the old guy sitting down surrounded by the remaining on duty crew at Station 15. He was ill, but not critically. Sitting down for a bit had made him feel better and he decided he didn't want to go to the hospital anymore. As Fat Head ran very few ambulance calls, an unnecessary one was enough to set him right off.

"You sure you don't want to go?" Fat Head questioned the old guy.

"Yeah, I'm sure."

"You're positive?"

"Yeah."

"Okay, then get your ass out there and wash that ambulance," Fat Head instructed.

"What?"

"You heard me. I got to wash that thing now that we came all the way up here, so if you feel so good you don't need to go to the hospital, you can wash it." Fat Head winked at the surrounding crew.

"You're kidding me, right?" the old man asked looking apprehensively at Fat Head.

"Do I look like I'm kidding?" Fat Head looked menacingly at him, trying desperately not to crack a smile.

"Okay, you win. I'll go."

They loaded the old guy up and transported him to the hospital, and no, Fat Head didn't wash the ambulance when the crew returned to the station.

• • •

A relaxing sunny spring Saturday afternoon was rudely interrupted by the bunk room and bay lights illuminating and the screech of the tones. A house fire assignment was dispatched for a home in a nearby residential development.

I was riding the left well or jump seat of the wagon. I grabbed the air pack out of the bracket and hoisted it onto the diamond plate dog house over the roaring diesel engine. Careful to keep my balance on the swaying speeding engine, I turned outward and slung the straps over my shoulders, tightening them in the same motion. I draped the face piece over my head and connected the hose to the regulator. Straps tightened, gloves ready, I held onto the bar across the top of the cab, facing forward watching for smoke as we sped toward the house.

The siren screamed and the air horn blasted, as the foot of Sergeant H. danced between the two pedals on the floor of the engine. The staccato air horn blasts interspersed between the rising and falling of the siren were an art; similar to a drummer working the bass pedal and high hats. Sergeant H. was one of my favorite officers to work with; he was a tall blonde guy with a walrus mustache, highly competent and obviously on his way to a higher rank.

We turned into the neighborhood, which was full of kids on bicycles and men washing cars and mowing grass. All were blissfully oblivious to the emergency facing one of their neighbors. That would end as soon as the streets filled with additional apparatus.

The engine and truck were both first due on the call. The driver brought the engine to a stop at an intersection, the truck still a half mile or so behind us. The window between the officer and the jump seats slid open, and Sergeant H. yelled back to us, "Lay out!"

The firefighter in the left well or jump seat, me in this case, was the designated layout man. I climbed down and trotted to the rear of the engine where I grabbed the Humat hydrant valve from its bracket along with the first length of three inch supply line in the hose bed. I pulled straight back toward the intersection and looked at each corner for the hydrant.

Uh oh, I couldn't see a hydrant. I glanced around each corner; nothing there, either.

I dropped the valve and hose in the middle of the street and ran up to the officer's side of the cab.

"There's no hydrant here, Sarge," I told him. He glanced down quickly at the map book on the rack in front of him, which showed, at least supposedly, all the hydrant locations.

"Split lay, split lay," he yelled down to me. That I could do. He would have to radio the next due engine where to pick up our line so they could lay from there to a hydrant. Their layout man would connect the two lines, and their driver would tie into the hydrant when they reached it.

I ran back to the idle valve and hose and picked them up. I glanced around, picked out a telephone pole on the nearby corner, and dragged the hose over to it. I wrapped the line around it, careful to place the valve under the hose, so the engine would not pull it loose, and then yelled back at the engine, "Go!"

The driver took off, hose peeling from the bed with a clank every few seconds as a hose butt banged off the blacktop in fifty foot increments. After making sure the end of the line wouldn't come loose, I trotted after the engine, air pack slapping against my back as I went. A couple of hundred feet down the road, the engine stopped in front of an ordinary looking two story house.

I reached the engine and helped the firefighter from the right well start to stretch the attack line. There was nothing showing from the outside, but this was standard procedure. Sergeant H. had already gone into the house. We stretched the line to the front door and then into the foyer at the base of the stairs.

Looking up the stairs, we could see smoke, so we knew we had something. I pushed my helmet back, and put on my face piece. Air began to flow into the mask as I pulled my nomex hood up over my head and then replaced my helmet.

Sergeant H. came part way down the stairs and waved for us to bring the line up. We proceeded up the stairs to the hallway and I flaked the line out behind us to give us some extra to advance more easily once it was charged with water.

There was a bedroom on our left with fire burning in the corner on the far side of the bed. It was burning up the wall and just starting to roll across the ceiling. A piece of cake.

We positioned ourselves in the bedroom doorway with the engine officer. He keyed the mic attached to the collar of his bunker coat and said, "Portable Engine 241 to Engine 241, charge the line." We waited. No response and no water. Again. "Portable Engine 241 to Engine 241, charge the line." Same results. The Sergeant was getting pissed, and the fire was growing. A third time. "Portable Engine 241 to Engine 241, charge the line." Again nothing.

Now the Sergeant was livid. He crawled into the front bedroom, which faced the road. We lost sight of him in the smoke, but we heard the window slam as he threw it open.

Then we, and everyone for three blocks, heard him.

"Hey!" he yelled down at the pump operator in a clear voice, having taken off his face piece. The pump operator looked up at him from the street. "Charge the fucking line!" That got the pump operator's attention, as well as the attention of most of the mothers and children in the crowd of spectators. We were more concerned about the pump operator at that point, but we knew it worked when we felt the line come alive.

Thirty seconds of water, and the fire was out. We spent another half hour or so overhauling and repacking the attack and supply lines, and then started back to quarters.

From the jump seat, I couldn't hear the conversation in the cab between Sergeant H. and the pump operator, but it appeared to be rather animated. Actually, it wasn't really a conversation, because two people have to talk to have one, and in this case, the officer was doing all the talking, and the driver was listening. I was happy not to be on the receiving end of that.

• • •

We were on our way down to Station 18 on a transfer. The Kensington area units were on a working fire in a garden apartment complex on Georgia Avenue. Both the engine and truck from Station 24 had been sent on the move up. We would stand by and cover the large area vacated by the units working at the fire.

The leisurely ride was interrupted by a long series of tones over the radio speaker above my head. The dispatcher announced "second alarm on box…" The red lights came on and the siren wailed under the heavy right foot of Calley, the technician in the officer's seat on the ladder truck I was riding.

I stood and held up two fingers so the tillerman could see. He had no communication with the cab and no radio speaker way back there at the other end of the ladder. The fingers told him we were going on the second alarm.

On arrival we were directed to one of the garden apartment buildings in the complex. Strangely, hose lines were already going into the front door of the neighboring building. As I passed the front windows of the building we were assigned to, I could see fire rolling in the bin room below grade.

All these garden apartments had bin rooms, which are a storage room subdivided with chicken wire walls for the apartment dwellers to store larger items like bikes, lawn furniture, and similar items they might not want to keep in their apartments.

A line was stretched in, and a guy from Truck 5 and I started up the stairs to search the apartment over the fire. Above the fire is the most dangerous place to be, for both victims and firefighters.

We forced the door and began a quick primary search of the apartment. It was hot and smoky, but not terribly so. The fire hadn't extended up there yet. The apartment was clean, and we went across the hall to the next unit. There was less smoke and heat, so we were able to cover that one more quickly. By the time we had searched both apartments, the engine crew had knocked the fire down. I went down to the bin room to help with the overhaul.

One of the chiefs came in and explained to us what had happened. We had been speculating on why the hell the second alarm units had the first line on the damn fire?

The first alarm units had arrived and gone to work on a bin room fire in the adjacent unit. While they knocked that down, the incident commander was seeing more smoke outside than he thought he should, so he took a short walk. When he got to the next building, he received quite a surprise: another bin room off and rocking.

We laughed as we pictured the poor chief's reaction when he found a burning building half a block away from where he had crews working. All his manpower already committed there, he hit the second alarm bringing us to the scene. Some arsonist had been busy that morning.

The chief started releasing units from the Georgia Avenue fire. We had a bunch of equipment to pick up before we could go back into service. As we started hauling stuff back to the truck, we could hear over the radio the beginnings of something big in Takoma Park.

House fire assignment after house fire assignment were being dispatched, all in the same general neighborhood. It sounded like a damn conflagration.

I started hurrying, practically throwing equipment back on the truck.

"You gonna bid on it, Calley?" I asked the truck officer hopefully. Bid was our term for trying to get yourself assigned to a call to which you would not otherwise be due to respond.

"Hell no," he answered. "I just wanna go back and watch some TV."

Calley had been around a long time and seen his share of fire. He'd go if sent, but he had no real desire to spend the rest of the shift in a rougher part of the county.

"C'mon, Calley," I practically begged.

"Listen kid, I'm pushing the button to put us back in, and hopefully they're so busy down there, they won't notice." Calley figured the simple change in color on the display terminal in the dispatch center from "on scene" to back "in service" would go unnoticed in the maelstrom of incoming calls.

I was disappointed, but knew better than to push the issue. I climbed back into my jump seat and we started back to the station. Just as he had said, Calley pushed the "in service" button on the modat head. I heard the standard turkey gobble noise over the radio as it transmitted. There was no response from the dispatcher. It looked like he would get his wish. My shoulders sagged with disappointment.

We were winding our way through the ever-present suburban D.C. traffic when the dispatcher finally noticed we were in service.

"Montgomery to Truck 24, are you available?" came the question. Now it was Calley's turn. I could see his shoulders sag in the cab as he picked up the mic. So much for television.

"Truck 24's ready," he casually answered.

"Truck 24, respond on the Takoma Park incident. Report to staging at Station 2," the dispatcher directed.

Yes! We'll make it to the big one.

After about twenty minutes, we pulled into Station 2. A guy I knew from Prince George's County Engine 41 had just gotten there as well. (Their engine had been one of the first units originally dispatched, but had been sent to the station for a break.) He walked up to me and kissed me on the forehead.

"It was fucking awesome," he started to explain. "We pulled in and stretched to the house, went in, and got a quick knock. We came out the door, and the fucking house across the street is going. We drag the line over there and got that one. Again we come out, and the house next door is fucking rockin.' Three fucking houses, and we never moved the engine."

I wish I could say we'd been sent out of staging and into the mess, but they had enough units there by then. We sat in staging for the rest of the morning, Calley getting more and more pissed off about missing his soap operas or whatever the hell he wanted to watch.

The explanation came later that day. A gasoline tanker driver had pulled into a local gas station to make a delivery. Somehow, instead of his hose going into the underground storage tank, it went into the storm sewer system. Gas fumes backed into the cellars of the homes in the area, and in sixteen cases, found an ignition source such as a furnace or gas fired water heater. Yes, sixteen houses within a few blocks burned.

It wasn't a conflagration but, as far as I was concerned, it was close enough.

• • •

Even on a weekend home with my parents, the fires didn't stop. "Gary, wake up, you've got a house fire!" My mother was hovering over me, trying to wake me up. "Huh?" I said, looking over at the Plectron. The light wasn't flashing and I sure as hell hadn't heard it go off.

"Where?" I asked, still half asleep.

"What difference does it make? You're going with your father. Get your bunker pants on!"

I was pretty out of it. I had driven home from school that day, and we had run three calls the night before, so I'd had very little sleep in the previous twenty-four hours.

I rolled out of bed and stuck my feet in my boots, pulling the pants up and straps over my shoulders in one well-practiced motion. That was one thing I could do while practically asleep. Mom hustled me down the stairs, repeating that I didn't need to know exactly where I was going, since I was going with Dad.

By the time I made it to the garage, I was a bit more awake. Dad explained that the Plectron hadn't gone off; the dispatcher had called him by phone. There was some sort of malfunction at the communications center in Endicott, and she was unable to set off the tones. As pretty much nobody listened to the open channel in the middle of the night, putting out the call voice-only was useless.

Before we left the house, Dad asked Mom to start making calls using the phone list taped inside the kitchen cupboard. "Start with the assistant chiefs, and then the guys closest to the station," he told her.

Dad backed the car out of the garage and I hit the switch for the light bar. I flicked the siren to yelp right in the driveway. There was no traffic at that time of night, but I hoped the siren would wake up Roy, across the street, and Doug, another guy down on the corner.

"Are we going to the scene or the station?" I asked.

"Station," he responded. "You and I aren't going to do much good on the scene by ourselves."

We got to the station and I grabbed the rest of my gear while Dad unlocked the door. He immediately went into the small electrical room just inside the door where the station siren controls were and began to blow the big unit by hand, pushing the button to bring it up and releasing it to let it drop in tone. It was the way firefighters were alerted in the old days when he started. Maybe it would still work. I dropped my gear and went to the radio room and tried to raise the county control center via the base radio.

I could hear the siren on the building wailing as I keyed the base station microphone. "West Corners Station 1 to Broome," I called. If I could raise them, they also had the capability to activate our receivers. I don't know why we hadn't told Endicott to call them; I can only plead exhaustion and the strangeness of the entire situation. Twice more I called, the third time was the charm. "Hit our tones, house fire, fire in the cellar, 818 Wallace Street," I told them. Without even waiting to hear back, I punched the buttons to open all the apparatus doors. Dad had heard the radio conversation over the station speakers and headed out the personnel door to get back in the car and to the scene.

I ran over to the white over red Seagrave Engine, jumped into the cab, and started it up, punched the switches for the red lights, and immediately realized that I didn't want to drive. The last thing I wanted to do was get stuck on the pump while there was a good fire to fight.

Jumping out of the cab, I ran over to the back of the squad and climbed in the rear. I put on my gear and then a pack from the interior wall of the rig. Peaking out of the rear door and around the side, I saw Dennis run in and jump in the cab of the engine to drive. *Beautiful.* Now I could ride the first piece and likely get the nozzle. I scooted back over and got in the right jump seat, the next few guys filling out the crew. Within a minute we were out the door, siren yelping in the cool night air. The house was only about two minutes away. We pulled up and I saw Dad waiting for us in the front yard. As I was the only one already packed up fully, I went around back to see how to get to the fire while the officer, Roy, and the other guy, finished donning their SCBA and stretched the line to the rear of the house.

The owner was a career guy from Endicott. When we arrived, the smoke was still banked nicely, and the owner told me to follow him to the fire. It was a good thing, because the basement was a damn maze. We went in, turned left, turned right, went straight, turned right, then left, and then left again to a doorway. *Damn, it was a fucking bomb shelter.*

In the 1950s, if you built a house, there apparently was some grant money available if you agreed to put a bomb shelter in it. Whoever had built this place, had obviously taken advantage of this Cold War program.

Back out we went, and the crew arrived with the line. "We'll take it in dry," I said through my face piece, knowing it would be a bitch to stretch a charged line around all those corners. Left, right, straight, left again to the doorway, the nozzle in my hand, as the line came to life. We started down the hallway and toward the entrance to the shelter proper. The fire rolled out this door into the concrete passage. I needed to get into that doorway to get the actual fire. About half way down, the guy behind me started screaming his neck was burning. We backed out and regrouped. We tried again; same result.

I ran back to the entry door and told Chick, the assistant chief in the rear of the house, we were having a hard time, and then went back to the line. It was really getting hot in the outer room. While on my knees, I raised my gloved hand above my head to try to gauge the heat. I didn't get too awful high up before I had to yank it back down. I told the captain I thought we had one more shot before the outer room flashed.

In this section of New York State, hoods had not yet caught on. We wore them in Montgomery County, though, and I had mine on that night. My theory has always been that you really shouldn't go any deeper into the fire with a hood on than you would without. It's designed for additional protection against unforeseen events, not to allow deeper penetration on a routine basis. In this situation, we didn't want the outer room to flash—the room and its contents all seemingly igniting at the same time—and possibly lose the whole place. I didn't have much choice.

We made a quick plan. I started down the passage on my stomach, the nozzle out in front of me, with no backup man. They kept things in the hallway clear behind me, and fed me line from the outer room. As I made it to the corner of the doorway, I opened the nozzle on straight stream and stuck it around the corner, angling it up toward the ceiling. The steam started, and the plume reversed, dumping on me. *Fuck, it was hot.* My bell went off, indicating my air was low. I needed six more inches. No way was I giving up that close to success. I humped forward one more time and was into the doorway. I rolled onto my side and blasted the entire room, whipping the line around. In thirty seconds, it was blacked. I shut down and backed out and told the captain it was knocked. I went out to get rid of my pack, totally wasted. I found Dad and told him I had made the room.

It was so hot in the shelter, it was close to an hour before we could get a crew in to fully overhaul the mess. It was packed with magazines, newspapers, and similar items. There was a heavy I-beam on the underside of the concrete roof of the shelter that was deformed into a C-shape by the high heat. Not exactly the use they expected for this room back when it was built during the days of Khrushchev and Sputnik.

• • •

While it was fun to visit home, especially since I caught some fire, it was back to school the next day. A few evenings later, we were hit for a call. "Trailer on fire with people trapped," the dispatcher's voice announced over the station speakers, the kind of run that gives everyone a double shot of adrenaline.

I rode the ladder truck, and even from a distance I could see the glow in the night sky. It was not a good sign.

We pulled in behind the engine. There was heavy fire showing from the kitchen and living room end of the trailer. It didn't appear that the fire had yet reached the end bedroom.

The engine crew pulled a hand line and attempted to hold the fire back a bit from this bedroom from the outside while we opened up the side wall with the K12 circular saw. Two quick cuts forming the legs of a triangle were made, and the skin of the trailer quickly folded down. A couple of pokes with the halligan bar, a heavy metal pry-bar with a pointed adze on one end, and we had the interior wall open.

I leaned into the hole with another guy from the truck and we searched the floor in front of us as far as we could reach. We found a mattress on the floor, but it was empty.

So far, the combination of the closed bedroom door and the hose line had kept the fire out of the bedroom. We backed out of the hole to put our face pieces on. We were going to try to get in through the hole to search the rest of the bedroom.

I finished putting my face piece on and pulled my hood over the straps when the bedroom door let go. The room flashed and fire blew out the hole we had cut. Not good news for anybody who might still have been in there.

The engine crew reacted swiftly, getting the line into the hole and pushing the fire back into the hallway. Timing is everything. If we had been in the room when that door let loose, it wouldn't have been pretty.

"Get that room searched," the chief told us. We crawled in the now smoldering room, as the engine guys covered us from the opening. We made a quick trip around the room.

I wasn't sure I wanted to actually find what would be left of anybody in there, but that was the job. The room was clean, though. We exited from the hole and gave our report to the chief.

With no possibility of a rescue left, the engine company started attacking the fire from the doorway at the other end. They entered and starting knocking it down, working their way through the trailer. We followed behind with our hooks and tools, searching and opening up as they went. As with most trailers, we made quick work of the fire, but there wasn't much left. At least there was good news—there were no victims.

With the fire pretty well knocked down, we concentrated on overhaul. With the floor was gone in places, we had to carefully step from joist to joist as we continued to hunt down hidden areas of fire.

The one officer on the call was, shall we say, less than popular. He didn't bother me at all, but some of the guys disliked him intensely. I stood next to the nozzle man, my tool in hand, looking for areas that night need work. The officer was about ten feet in front of us, looking at something else.

The nozzle man whispered to me, "Fall on me." I gave him a quizzical look, not sure I heard him correctly. He winked and repeated his strange request for me to fall on him.

We were both standing on nothing but floor joists, so falling wouldn't be anything unusual. Still not sure what was going on, but willing to play along, I leaned sideways and pretended to lose my balance, grabbing hold of the nozzleman to steady myself.

"Whoa," I yelled.

At that point, the nozzleman yanked the bale back, and having carefully aimed, hit the officer square in the chest with a straight stream, a hundred fifty gallons per minute, knocking him on his ass.

Closing the nozzle, he grabbed me and with great solicitude, checked on my well-being.

"Are you okay?" he asked me. "Sorry Sarge," he called out to the officer as he picked himself up off the floor, glaring at the nozzleman.

"He fell on me and the nob opened by accident. Are you okay, Sarge?" The nozzleman continued to apologize profusely, but the smile on his face made it clear he was full of shit. The officer just continued to glare, and after regaining his feet, made his way past the two of us and out of the building, not saying a word.

Within minutes of returning to the station, everyone knew about my little "fall," except the officers, that is. In reality they must have known, too. Most of them didn't care much for that particular sergeant, either. *Hey, accidents happen.*

• CHAPTER TEN •

Firefighting 101

The career officers in Montgomery County were great and kept a quasi-parental eye on their young charges as they understood that our primary purpose was to go to school. It happened rarely—and never to me—but if they did notice one of the college guys spending a bit too much time at the station, not attending class or studying, they'd give him a kick in the ass. If it got so far out of hand that they needed a second reminder, it was probably too late. One kick was usually enough.

My social life was a little more limited than it would have been if I had lived on campus, but the bunkers still occasionally went out to bars or for a pizza. I lived on a strict budget and, therefore, became adept at negotiating with the girl I was dating.

"If you want to go out Saturday night, I need three dinners at your house this week," I would say. She would smile and procure the necessary invitations from her mother. It was definitely better than McDonald's or eating on campus and less stressful than eating at the station where interruptions during meals were commonplace.

The station was a terrific place to live, particularly for a college student. Free room and cheap board—I could buy into the meals with the career guys if I wanted—made it more affordable than dorm living. Studying Fire Science at the University of Maryland was a great experience, thanks to the first rate professors, but maintaining the balance between school and firefighting was a challenge. It was good to be young, as I slept eight hours straight only about one night a week. The rest of the nights were either interrupted by calls or by late night bullshit sessions.

Weekends were, sometimes, more relaxing. Our Sunday morning ritual always consisted of a cup of coffee, a glance at the newspaper, and watching our regular dose of a televangelist who believed he had healing powers. He would pop some poor soul on the forehead with the heel of his hand and scream, "Be healed!" at the top of his lungs. The person would drop to the ground. When they got up, they were cured of whatever had ailed them. Although this show was not meant to be a comedy, we found it hysterical, and the laughter and imitations of the faith healer went on throughout the program. This was done in the typical firehouse vocabulary.

"Cocksucker" and "motherfucker" were utilized in virtually every sentence. The dayroom was not an environment for the faint of heart.

That morning, some poor woman had, unbeknownst to us, entered the building and made her way halfway down the hallway to the day room. When she got close enough that she could hear our conversation, she stopped, and proceeded no further. Only the critical news she needed to convey kept her in the building.

One of the guys left to hit the head and saw the poor woman standing in the hallway.

"Can I help you, ma'am?" he inquired. Probably the first sentence he'd uttered all morning without the word "fuck" in it.

She gestured down the street.

"Th...th...there's a house on fire around the corner," she stuttered.

He didn't answer her, but yelled back to us, "Hey, we got one."

The poor lady was almost trampled in the hallway as the foul-mouthed crew headed for the apparatus floor.

The fire turned out to be a pretty routine chimney fire, and we were glad she stopped and told us about it. Too bad we had to miss some of our favorite show, though.

• • •

Pelcie, one of the truck officers, was short and stout, with a jutting jaw and the appetite and eating habits of a pelican, hence his nickname. He had a high voice, which was typically used at a substantial decibel level. He was an expert at forcible entry, especially in the less destructive "through the lock" method. He was rude, crude, and opinionated, the exact opposite of his best friend, Doc, a paramedic and Renaissance man who worked the medic unit at the station down the road. Pelcie was a pain in the ass to the platoon and station officers, but a consummate professional when we went out the door. And he was the best truck officer I've ever seen.

One day, we were dispatched with Engine 151 on a gas leak. Originally, it had been a single unit dispatch, since these were typically routine three quarter inch residential lines. This call was a tad different. A huge office building under construction had a six inch gas main cut near the building foundation. On arrival, Engine 151 filled the box, having a full structure fire assignment dispatched, which brought our crew on Truck 24 as well as multiple other units.

We could hear the massive quantities of gas screaming from the ruptured pipe when we arrived. The crew from Engine 151 had a master stream set up flowing water to disperse the gas so that it couldn't accumulate and ignite. We were tasked with searching the building for low spots or similar areas where gas pockets could form.

Over the radio, the announcement was repeated every couple of minutes, "All units on the fire ground, Route 29, use of SCBA is required."

Don and I were following Pelcie around the building. Our SCBA, kind of like scuba without the "u," were on our backs, but the face pieces themselves hung loosely around our necks. The announcement was repeated and Don and I glanced at each other uncomfortably. I gestured at him and him at me. Neither one of us wanted to be the one to question Pelcie.

Finally, Don lost, or won, depending on how you look at it.

"Uh, Pelcie, did you hear the announcement? Shouldn't we be on air?" he asked.

Pelcie froze and slowly turned to face us. His eyes narrowed as he looked at his two young charges.

"What's our job in here?" he asked us.

"Look for gas pockets," we both answered.

The explosion then came: "How the fuck are you going to smell the gas if you have a face piece on!"

Satisfied, he turned on his heel and continued on. We looked at each other and shrugged at his common sense explanation. Orders don't always make complete sense.

• • •

Shortly after lunch on a routine weekday, a box alarm sounded for a fire at the high rise apartment building at 11700 Old Columbia Pike, one of our regular destinations. This twenty-two story building produced its share of serious fires over the years. Engines 121, 241, 161, 551, Trucks 24 and 16 were due.

Arriving on the fire floor, the odor of burnt food on the stove greeted our noses. A couple of the engine guys moved the smoking pot from the stove into the sink and ran some water in it. Those of us on the truck got some windows open to ventilate. All other companies on the box were told to go back in service.

About this time, the apartment's occupant returned home, somewhat unnerved to find a bunch of tool-festooned firefighters crawling all over the place. She owned one of those yippy dogs—a dog so small that it defies a breed name and has a high-pitched, annoying yip that is out of scale in comparison to its small size. This one was incredibly happy to see its owner.

The dog began jumping on Mommy's leg, trying unsuccessfully to get her attention, which was distracted by the recent invasion of her apartment by the yellow-coated monsters. She put her purse down by the edge of the couch, continuing to ignore the near-frantic animal. The dog jumped onto the arm of the couch, apparently thinking a little elevation might increase its attention getting ability. The yipping continued. Mommy still didn't respond. When this didn't work, the dog had one more method to try to get Mommy's attention—it shit down the side of the couch, right into her owner's waiting purse!

The laughter in the apartment was uproarious. The dog's owner was furious. Not at the dog, for some reason, but at us for laughing. We picked up our equipment and left, the woman's now loud bitching following us down the corridor as we continued to laugh and shake our heads.

• • •

Our operations in the field were usually very professional and disciplined. You wouldn't even think of referring to the engine or truck officer by his first name while on a call; only by their title, and I wouldn't even speak to a chief unless spoken to first. As professional as things were typically handled on calls, occasionally things could go over the line in the station.

Don was a great firefighter. He was a live-in who later went on to a career firefighter position. He was also a bit of a masochist. He enjoyed being abused by the career guys. He would pick and prod and irritate until one or more of them would throw him on the couch and rabbit punch him into submission. One day, however, things went a bit too far.

It was a weekend day shift and the technician was the acting officer. As usual, Don picked away trying to get a rise out of the guys and, before long, he had the whole platoon fired up beyond his wildest dreams. A bunch of guys threw him on the couch and punched the hell out of him. But it didn't end there; somebody pulled his pants down while someone else grabbed a handful of ice out of the freezer. They held him down and threw the ice on his crotch. Needless to say, he found this a bit uncomfortable, and bucked the ice off like a bronco tossing a cowboy.

Pelcie pulled a knife from his pocket and opened the blade. He picked up one of Don's balls by a pubic hair and put the knife underneath, releasing his testicles to rest on the blade. He then placed the ice back in his crotch and told him, "Now move, motherfucker."

Don didn't sustain any injuries; but things were interesting during the shift change the next morning. The incoming officer had somehow found out what happened, and proceeded to dress down the out-going platoon. It was the most incredible ass-chewing I had ever seen in my entire life. I learned multi-word combinations using "fuck" that I had no idea existed. I was so impressed, I forgot to be shocked. Things were much calmer around the station after this...for a day or two.

Don's dysfunctional relationship extended to all three platoons in the station, not just one. Around the time of the Falklands invasion, one of the other platoons became intrigued by the news accounts of the ongoing battles between Britain and Argentina. They were enthralled by it, like a group of kids playing war or soldier.

Don, as usual, was being his typical irritating self. Pretending they were Argentinean terrorists, members of the platoon kidnapped him, tied his hands together, and placed a pillowcase over his head. They then dragged him into the hose tower, attached the hook to the back of his belt, and raised him up in the tower where they left him dangling.

On another occasion, customers of the drive-through lane at the bank next to the station were treated to an interesting sight. It was Don, stripped to his underwear and tied to a chair, which was secured to one of the pillars supporting the drive through roof.

It was no wonder the station was known as the "animal house"...

• • •

A frazzled old drunk with a million miles on him knocked on the front door of the station one night, his arms weighed down with a dripping, briny case of oysters. Our ambulance had treated him and transported him to the hospital a few days before, and he was grateful for the care he had received. As his way of saying thanks, he gave the crew on duty the case of oysters, which he had somehow acquired.

Mouths watering, the career staff put together a short shopping list for a sauce for the oysters and sent me to the nearby grocery store. Upon my return, a huge kettle was beginning to boil, and the oysters were added to steam. When the first batch and the sauce were ready, the crew began an orgiastic consumption. Under threat of death, I tried one of the mollusks. A boiled rubber band would have tasted better to me. I was perfectly content to retreat to the television and leave them to their eating and moaning about how good they were.

Shortly into their meal, the station alarm sounded. It was not an emergency, but a dispatch to transfer the engine to Station 15 while they were operating on a working fire in Howard County.

I walked out onto the apparatus floor and over to the engine. I put on my coat and helmet, and threw my bunker pants up into one of the jump seats. I climbed up and secured the chrome safety bar.

Normally by this time, the bay door would have been open, and I would have had plenty of company manning the apparatus. I began to wonder if I had missed something, like the call getting cancelled. About this time the station bell rang, with the dispatcher calling on the direct line. I waited another half minute or so, and still no one else joined me. I assumed I had made a mistake and started to climb off the engine.

Suddenly, I heard footsteps running in my direction. The swinging doors from the hallway burst open with a crash, and I saw three of the career staff come around the front of the ladder truck. Their arms were filled with pots full of sauce and the huge oyster kettle. The bay door opened while the pots and pans were shoved into the cab. The engine came to life as the rest of the crew jumped on and we began our transit up to the Burtonsville station. After a short ride, we arrived and backed into an empty bay. The pots and pans were quickly removed from the cab and placed on the stove in the Station 15 kitchen, where the eating continued.

• • •

While this area was normally known for mild winters, occasionally Mother Nature could throw a wild party. In February of 1983, the southern section of Maryland and the District of Columbia got hit with a whopper snowstorm. Eighteen inches of the white stuff fell, paralyzing the entire metropolitan area. Folks there don't like to drive in it, don't know how to drive in it, and don't have the equipment to get rid of it.

For some reason I ended up assigned to the engine that night. It figured; the truck didn't turn a wheel. The engine, on the other hand, ran eighteen calls in twenty-four hours. We barely had time to breathe.

At least half the calls were assisting the ambulance on maternity runs. It seemed like every pregnant woman in our area who was six months along or more had come to the same conclusion: the storm was the end of the world, and there was no possible way the snow would be cleared before they actually went into labor. Therefore, they wanted to go to the hospital now, so they would be there when the baby decided it was time. There was no debating any of this with them. They all went to the hospital, and by the next day, since Maryland in mid-February can be twenty degrees one day and sixty the next, they all went home.

• • •

Running lots of calls was great, but I was really there to study. Even getting out of my station didn't guarantee time with the books. I liked to study in the basement room at nearby Station 12, which was set aside for the college guys there. The nice part about studying there was their top-of-the-line IBM Selectric typewriter—considered high technology back then—which was great for reports and papers. The other nice part, at least for me, was that because I wasn't a live-in at Station 12, I didn't have to go on every response. I could cherry-pick the calls and go on what sounded like the more interesting runs.

My studying was interrupted by a call for a wreck on Route 495, the Capital Beltway. These calls could always be interesting, so I decided to tag along. I ran upstairs and grabbed some gear off the rack, and climbed into the left well of the wagon, along with another college guy. Mary, a young, female career fire-fighter was in the right well, and Lieutenant R. had the seat.

Left onto the avenue and then up onto the Beltway we went, following the medic unit, Medic 4. On arrival, it was pretty obvious the wreck was minor, and there really wasn't much of anything for us to do there. The paramedics loaded the driver, who had minor injuries, and we loaded ourselves back onto the engine and started back to quarters.

I began to think I should have stayed in the basement studying, when the county started putting out a box for a building fire in a complex of townhouses. Even better, we were the first due engine.

The lights came on, the siren started screaming, and I grabbed the MSA air pack, and slung it onto my back. The other college guy and I started to discuss whether we would lay out or not. With every other officer we worked with, except Lieutenant R., if a full box or house fire assignment was hit, you laid a supply line, no questions asked. With him, you never knew. It seemed like he had a sixth sense or something. We could have a call that didn't sound like much, nothing visible, and he would drop a line, and damned if we wouldn't have a fire. Conversely, we'd get hit for something that sounded like the end of the world, and he'd have the driver blow right past the hydrant and again be right; we'd have nothing.

So with Lieutenant R., the first indication on whether we had something or not would be whether we laid a supply line. We turned into the complex. Scanning the sky, I couldn't see any smoke, but when we pulled up to the plug, the wagon stopped, and his slow baritone drawl called out through the sliding window, "Lay out!"

My partner jumped off, grabbed the Humat hydrant valve and first length of hose and wrapped the hydrant. The driver continued down the roadway, three inch hose peeling off the bed behind us. We pulled up to a nondescript townhouse—like a couple hundred others in the complex—with nothing showing.

With most other officers, we'd automatically stretch an inch and three quarter line to the doorway with our tools, a Halligan, an axe, and a K-tool, but not with Lieutenant R. With him, we'd only do something once told, so we dismounted and stood by the side of the engine awaiting his orders. With just his coat and helmet on, no bunker pants or SCBA, Lieutenant R. strolled into the front door of the building in question. A minute later, he walked back out and over to us, at the same casual pace he had walked into the place. When he got back to the three of us, he nonchalantly drawled, "There's a room on fire on the second floor. Go put her out."

Young pups that we were, we grabbed tools and the hose line, quickly stretching a pre-connect to the front door, pausing to don our face pieces and pull up our hoods. Mary took the nozzle with me and the other college guy backing her up. We advanced the charged line up the stairs. The room on the left was off, but it hadn't extended into the hallway. Mary turned into the doorway and opened the nozzle at the ceiling. Thirty seconds of water and we pushed into the room, the fire essentially out.

The truck company, my truck, wasn't on scene yet, so we didn't have any ventilation and the smoke was still pretty thick. I crawled across the hallway into the front bedroom to get the windows open, hoping the smoke would start lifting. After opening the window in that room to little affect, I crawled back into the fire room—bang!—my helmet ran into something. I was an over-confident dumb ass and had left the security of the wall thinking I knew where the door was. I found the door all right, the one that led me directly into a closet.

Okay, time to regroup and figure this out. I backed out, reacquired the wall, and found the door into the hall.

About this time, the first due truck and second due engine personnel were arriving, so we could take a breather. We walked down the stairs and outside, and I took off my helmet, hood, and face piece. When the guys from Station 24 saw it was me, they started busting my balls. *You traitor, how could you run with Station 12?* It was good-natured, but they were a little envious I got to be first due.

• • •

Another study session led to an important lesson. I rode the left jump seat on Engine 121 responding to a tractor trailer/car accident on the Beltway. As we proceeded down New Hampshire Avenue we could see the wreck on the overpass in the distance. The car was jammed under the trailer with what appeared to be serious damage. Back then, jump seats weren't enclosed, and we rode them standing up facing forward. The career guy in the right jump seat reached down and grabbed his air pack and slung it on his back, preparing to don the face piece. I gave him a quizzical look.

He hollered over the sound of the siren, "Do you know what's in that truck?"

"No," I answered, still confused.

"Neither do I," he called back. "And we have to go right under it."

Oh shit. Now understanding the possibility of hazardous materials, I grabbed for my own air pack. There was no problem on that call, but it was a great lesson.

• • •

Another one I learned was that some folks don't understand English unless it includes regular and repeated use of the f-word. Early one morning, I sat half asleep in a jump seat on the ladder truck in the roadway of a large garden apartment complex. We were there for a reported gas leak. My yellow coat pulled tightly around me against the pre-dawn cold, I waited as our officer along with the officer from the engine investigated inside the building.

I vaguely noticed a well-dressed man exit one of the apartments and get into a parked car near the side of the ladder truck. He wasn't going anywhere, as we were stopped such that he was blocked in. That didn't prevent him from trying, however. He started his car and began to edge backward, trying to turn and squeeze along side the rig.

I watched through half open eyes, mildly curious as to how long it would take him to figure out he couldn't get out of his parking place. The other alternative, unthinkable, was hitting the truck.

Calley, the truck driver this night, thought it thinkable and deciding he didn't want to do the piles of paperwork necessary if this guy cracked the side of the rig, got out to talk to him. It was an illuminating conversation.

"Sir, please pull back into your parking place for a few minutes. We'll be leaving shortly."

"You don't understand, man. I gotta get back to the district," he said, referencing neighboring Washington D.C.

"Yes sir, I do understand, just hold off for a few minutes and we'll be moving."

His back and forth jockeying of the car continued unabated as he repeated, "I gotta get back to the district, man."

Both his fancy car and deportment at this ungodly hour screamed drug dealer. Too fancy for the regular folks that lived in this area but a step or two below pimp. Finally, Calley had his fill of this guy and put it in words he could understand.

"Motherfucker, put the fucking car in the fucking parking place and shut it the fuck off until we leave!"

Now understanding, the response was instantaneous. He pulled the car back into the space and the engine shut off. Calley walked back over to the truck and climbed in muttering the only pejorative that really meant something—*asshole*.

· CHAPTER ELEVEN ·

Pennsylvania Bound

After graduating from college in 1983, I worked a few part time gigs for about a year and a half, teaching first aid and CPR and working shifts as a fire technician for a large computer company, back home in Endicott. Eventually, I got a real job as a fire protection engineer in early 1985, and moved to a beautiful rural area in Northeastern Pennsylvania in 1986. Naturally, one of the first things I did was put in an application at the local fire department in Scott Township.

A week or two after I applied, I received a call to meet with the membership committee, which consisted of about six firefighters. When we all sat down together, Nicky, one of the assistant chiefs, asked me about my background and experience, so I told them a little bit about myself. Nicky was not a big guy, but was built like a plow horse, which was not surprising, since he came from a long line of farmers.

"Are you sure you live here?" Nicky asked with an ironic smile on his face. "'Cause we get people like you walking in the door just about every day." I assured him that I did actually live nearby and, with that, my probationary year began.

There were a bunch of fires and wrecks that first year. I learned the apparatus, the guys, and a bit about rural firefighting. It was quite a change to be someplace where there were no yellow things sticking out of the ground connected to large water mains. Beyond the lack of hydrants, there were other differences, too. Dirt roads, some well rutted, are common in Northeastern Pennsylvania, and lesser-used roads, mainly private, have grass growing between the tire tracks. These were not the wide paved streets I was used to. Street and road signs are provided haphazardly at best. Distances can be surprising. The first due engine could have a seven or eight minute run; the second due engine double that...once people get to the station.

After a year, my probation was up and I was voted in as a regular member of the department. Because of some personnel changes—not uncommon in volunteer departments—there was also an assistant chief opening, which I was quickly nominated to fill.

It wasn't as big a deal as it may seem. There were no captains or lieutenants there at the time, so my third assistant chief slot was the lowest officer in the department. It did come with a white helmet, though, and an opportunity, while still in my mid-twenties, to run some fires on my own.

. . .

Shortly after putting on the white helmet, I was the only officer around when the call came in for a house fire. As we pulled up, we saw smoke showing from a single story ranch house. From the officer's seat, I could see the lazy brown-grey fog beginning to fill the front yard. I already had extra help coming due to the early report of smoke from the building from our photographer, Henry. Henry was a tall skinny guy who had little taste for getting dirty. His gear was the cleanest of anyone in the department. Henry wore out more gear washing and drying it at the first sign of a speck of soot than he did from actual use. While he didn't have any interest in physically fighting a fire, he was an incredible photographer and helped out in other ways.

Getting men in and apparatus out during the day could be problematic, and so far, we had just five of us for the engine and tanker, plus Henry.

Sure, I'd run a bunch of calls—brush fires, car fires, wrecks, stuff like that—since putting the white helmet on, but this was going to be my first structure fire in command.

An inch and three quarter line was stretched to the front door while I did a quick three-sixty to size up the fire. It looked like it was in the left front corner bedroom, today the A/B corner of the house. Structures are all described the same way. The street or address side is the A side, and B, C, and D go clockwise around the building. I told the crew on the line to go in on the left and that I'd vent the window once they were in a few feet. They started in, and after watching another ten feet or so of hose disappear into the house, I took out the bedroom window with the side of an axe. The smoke started pushing and lifting, so I hoped their visibility improved.

Sonny, the chief from our sister department over the hill, arrived and walked up to me; he stood there in his work clothes, dark blue with his name over the breast pocket. He was a bit shorter than me, and looked like Radar O'Reilly with a beard. Sonny owned a garage, which he had left to come help us. "What do you need?" he asked me.

"Water," I responded. I was brand new at this rural water supply stuff and didn't have a clue. "I don't care where or how, just get me some water."

He smiled, nodded, and walked off. Sonny had been doing this forever, and I knew he could get a thousand gallons a minute out of a mud puddle. Sonny's knowledge of which ponds to use and, just as importantly, which ones not to; how to set up the draft engine and tanker fill site, and all the details that went with it, made him a great resource and teacher. If things went the way they should, none of Sonny's water would be needed, but a good chief always plans for things to go wrong.

The first line appeared to be making good progress, but I suspected some fire in the attic. I had a back-up line stretched as a precaution once some additional manpower arrived and told the truck crew to open the roof. I had called for a ladder truck out of the valley to assist; routine for me, but very out of the ordinary at the time for this rural department. After a few minutes, it started to look like the fire was pretty well knocked down.

About this time, Steve, the company chief, arrived and assumed command, and I told him we needed to finish getting the roof open as I left to go inside and run the interior, primarily for mop up. Over the year that I had been in the department, Steve and I had become good friends. He had dark hair—that the chief's job and his two daughters—ultimately helped turn gray, and was thin as a rail; so thin that he actually got a better mask seal with a beard because it helped fill out his face. While I was confident we had a good knock on the fire, I wasn't disappointed when he showed up.

Once inside, I could see the initial line had made a good stop. Fifteen minutes later we knew things were pretty well set. There had been some extension to the attic and a bathroom adjacent to the bedroom. The end verdict was a kid playing with matches. For me, it was a good start, though. *My first working fire as a chief and I didn't burn it down!*

• • •

Fighting fire in the country can present some interesting complications. Sometimes just figuring out where you're going can be a problem. Roads can have different names to different people, particularly to old-timers. One local road, Route 524, is also known by some people as Kennedy Creek Road or just Hill Road. Occasionally you can get dispatched to the old so-and-so farm. The so-and-so family may have owned that particular piece of property four generations ago, but to the caller that's what its name will be forever. You could get calls for a fire on "Old Swamp Stomper Road"—three houses past the barn that burned down in 1942. *Yeah, I know where that is, sure.*

When I became assistant chief, I had only lived here just over a year, and didn't know many of the back roads and special names for intersections, neighborhoods, etc. When a call would come in, Nicky, still an assistant chief, would stay at his house, which was right around the corner from the station, until he heard me call out responding.

He would then call me on the radio and inquire, "Do you know where that is?"

"Not a clue," was my usual response.

He would proceed to give me directions I could understand. Interesting way of learning a township.

• • •

Mutual aid calls to neighboring townships could be interesting as well.

The vehicle repair garage was well involved, flames boiling from the open bay doors. Originally we were transferred to Greenfield's station to stand by, but shortly after we arrived there, we were dispatched to the scene. They wanted our deck gun, the pre-plumbed nozzle on the top of the engine which could flow somewhere north of five hundred gallons per minute, something not available on the older Mack they were running at the time.

Positioned in front of the building, they got a supply line into our engine, and we gave the fire a good hard whack with the gun. The heavy straight stream darkened down the large volume of fire in the repair bays.

With that set up and operating properly, I took a walk around the building to get a better idea of the overall situation, mainly just being nosey. The rear of the building was actually partly below grade the way it was built into the bank. There was a row of interconnected rooms along this back wall that were also rocking, fire coming from the windows, but the gun and other lines in the front couldn't reach that far.

I continued my three-sixty tour of the building until I reached the front again, and found Greg, the Greenfield assistant chief who was running the fire. Greg was a good guy, very calm, methodical, and level-headed. With the big lines in front knocking down most of the fire in the bays, I wanted to try to do something with the rooms in the back.

"If we can get a line and an attic ladder back there, I think we can get those rooms," I told him and Ted, the other Greenfield assistant chief. Greg looked at me quizzically.

"We'll knock down the end room from the outside and stick the attic ladder in and take the line in over the ladder and go right down the row," I explained. Greg looked doubtful. Ted didn't, instead he volunteered to go. Greg acquiesced and, that settled it. Ted and I proceeded to get the ladder and the hose line in place.

Today, I wouldn't touch an operation like this without a second line with a full crew, additional manpower on the first line, discussion of a secondary route out, and similar safeguards, due to the limited access and below grade area we were dealing with. Ted and I were both young and stupid back then, still in our twenties, so none of that came to mind. Our only concession to common sense was to have somebody outside to hump the line to us. However, our choice of personnel could have been better.

The EMS weenie, who knew nothing about fire and was afraid to get dirty (his four- or five-year-old turn out gear looked like it had just come out of the box) was going to feed us the line.

When everything was set, Ted popped the fire in the first room and knocked it down. I slid the ladder in, the grey, shitty smoke still pouring from the window. Down we went, as quickly as we could to get below the heat near the ceiling, where the window was. It wasn't too bad near the floor. Ted took the nozzle and we started down the row of rooms. The first two were a piece of cake. I pulled the line along with us and leaned forward on it, taking the back pressure from the flowing nozzle off him. Then everything stopped, with one room to go.

"More line!" we both yelled through our face pieces. Nothing. We pulled on the line with all our might. Not an inch. *Fuck.* Ted stayed on the nozzle to hold the fire, and I followed the line back to yell at the moron ambulance officer who was supposed to be feeding us line. I got back to the base of the ladder all ready to tell him what a stupid jerk he was but he was gone. It must have been time for his coffee break. I tried pulling the line down from there; still no good. Looking through the doorway into the repair bays at the front of the building, I could see light through the debris.

Maybe there was an easier way instead of climbing out the attic ladder and pulling more line in myself. I worked my way through the debris, appearing out of the steam to the crews in front of the building, who were rather surprised to see someone walk out. I took off my helmet and pulled up my face piece and yelled to the assemblage.

"More line!" I pulled the face piece back down, put the helmet back on, and walked back through the smoke and steam to the rear rooms. By the time I got back, there were three or four guys outside ready to help push the line in.

A few more feet, thirty seconds of water, and we were done. Ted and I walked out front for a break, both still commenting on the "worthless fucker" who had abandoned us. He was lucky we couldn't find him right then. His medical skills might have come in quite handy, assuming he could practice them on himself.

• • •

It was a Friday happy hour, the start of the weekend. It was just beginning to get dark on a windy spring evening. My friend and fellow firefighter Alec and I had pulled our cars into the parking lot of a local establishment seconds before. We were looking forward to unwinding a little bit at the end of the workday, and I had some time to kill before the date I'd planned with Michelle, who was my girlfriend at the time. Before we could even get out of our cars, the pagers on our hips began beeping. The dispatch was for a structure fire on Matechak Road. I wasn't entirely sure where that was, still not familiar with all the back roads. Our best bet was to ride a piece out of the station. Unfortunately, Michelle would probably have to wait.

It was a straight shot down Route 438 to the station. I threw a red light onto the roof to clear traffic out of our way. About two minutes later, we pulled into the station, our tires spinning in the gravel. Cars don't really get parked in such a situation; more like abandoned. Trunks are thrown open to pull out bunker gear while we simultaneously kick off our shoes. As we pulled on our gear, the sounds of diesel engines firing, doors slamming, and feet running, were building to a crescendo in a fire station opera; it was music to my ears.

I headed to the officer's seat of the engine and climbed in. Tossing my white helmet on the dash, I reached back and put my arms through the shoulder straps of the air pack next to me. The rotating and flashing red lights from the trucks shot out the bay doors into the dusk like lasers. I looked back and checked the jump seats. Alec and Artie had climbed in and were settled in place, starting to put on their air packs. Old Todd, the driver, looked back as well, and then over at me.

"Go," I told him.

We went out the door like we were being launched off a carrier flight deck. I barely had time to reach over and flick the siren on before we were heading down the two lane country road.

I picked up the radio microphone and keyed it, "Scott Engine 4 responding." The tanker and rescue were behind us, and the other companies on the box were signing on with their apparatus as well.

We were heading back down Route 438 the way Alec and I had just come. I had no idea where we were going, but Todd knew the township like the back of his hand. Todd also knew only two positions with the accelerator, it seemed, but I could tell his favorite was all the way down. We slowed, barely, to make the right turn onto a small dirt road I'd never been on before. I saw a small one-lane bridge in front of us and wondered briefly about its rated weight limit. No matter—we never even touched it.

There was a small incline just before the bridge, and Todd used it like a ski jump. I'm not sure how long the heavy pumper was airborne but it was a hell of a lot longer than I wanted it to be. The chassis seemed to twist while in mid air. I'm no mechanic, but I don't think they were designed to do that. We slammed back down like on the flight deck of an aircraft carrier, except in this case, there was no arresting cable. We just kept on going. I was amazed the engine didn't split in half.

I turned around to see what I was sure would be two empty jump seats. No way anyone could have stayed in those tiny open slots. But Alec and Artie were still there, hanging on for dear life. Maybe they were more used to riding with Old Todd than I was.

I could see the red glow in the sky in front of us as we continued our flight down the dark country road. We reached the driveway to the old farmhouse and mercifully slowed to pull in.

The house was fully involved, transparent with fire. The wind had picked up and was blowing so hard that the flames through the roof were making a ninety degree turn; the fire in the sky burning sideways. There was a pop and what looked like a firework display lit off next to the house as the electrical service let go. It would've been beautiful in other circumstances.

The fire was exposing a barn, a couple of sheds, and a trailer. The three of us started stretching a heavy two and a half hose line to the rear to try to protect the buildings while Todd got us water. Over the stone wall of the farmyard we went with the line and got it in position. By the time we got water, one of the sheds was burning. We made quick work of that, wetting down the sides of the other exposed structures to cool them before turning the line on the burning farmhouse. Even with the big two and a half and two hundred fifty gallons of water per minute, it was like throwing snowballs at the sun. As more lines came into service, it started to darken down a bit, and the radiant heat was reduced, cutting the danger to the exposed buildings. We weren't going to put this fire out with the flow we had. It was going to have to burn down to the point where we could ultimately control it with the water we were putting on it.

We needed to relocate the line to be more effective. I went back to the stone wall to pull some additional hose over, which was easier said than done with the hundreds of pounds of water filling the line. I stood on one side of the wall, pulling until I thought my eyeballs would pop, gaining only inches at a time while, on the other side of the wall, Henry, the company photographer, stood calmly snapping my picture. I glared at him as he took the viewfinder down from his eye. He stood motionless, just watching me sweat. I snapped.

"Henry, if you don't put that camera down and start humping that fucking line, I'm going to stick it up your ass sideways," I told him. His eyes got wide, and his mouth opened and closed a couple of times, but nothing came out. He did, however, immediately grab the hose and start to push it over the wall while I pulled from my side. Hours of ass-busting work followed before the fire was out and all the equipment and hose were picked up. The remnants of the building were smoking in the fieldstone foundation.

Later, Henry and I talked about the little incident.

"When Artie was chief, he told me if I joined I could just take pictures," Henry said.

"I'll let you in on a secret, Henry; he lied to you. If you wear that gear, you work," I told him.

I missed my pizza dinner with Michelle. One important lesson was learned, though—try to ride whatever piece Todd wasn't driving.

• • •

Dating a firefighter was a new experience for Michelle. We were introduced by Alec, my fellow firefighter who was dating one of Michelle's friends at the time. I knew I liked her from the start, but it took her awhile to come around. After our first blind date, I stopped at the bank where she worked to try and see her. She was on break, and after looking around a bit, I left. It was a good thing I left; apparently one of the other tellers thought I was casing the place for a hold-up. *Not an auspicious beginning.*

Occasionally, I would be late or not show up at all for one of our dates because of a fire. A number of times the unwelcome beeping of my pager would interrupt the dates we did go on. I learned what dagger eyes were the first—and last—time my pager went off when we were in a movie theater. Through it all, though, she was incredibly tolerant and understanding.

When we got married, it was a major adjustment for "Shell," as I usually call her, to move from a neat, manicured suburban environment to the country. What struck her most was how damn dark it gets without streetlights and porch lights from neighbors who are right next door. Not to mention the fact that you can't get to the grocery store in two minutes if you run out of milk or eggs. Another new experience for her was meeting the critters; she liked seeing the occasional deer stroll through the yard, but the occasional field mouse that stops by still makes her a little nervous. There's a certain shriek I've come to know well that means a mouse has startled her as she walked into the kitchen to start the morning coffee. Even with all that, she now enjoys rural living.

After more than twenty years of marriage, Michelle still doesn't want to hear the end result of calls. Instead, she resets the pager and turns off the scanner so she doesn't hear anything that's happening. When I come home from a bad call, if I tell her anything at all, it's edited down for family viewing. Now, as a fire-fighter's mom, her exposure to calls has increased, and Mike hasn't learned to limit what he tells her the way I do.

• • •

I was finishing a bowl of ice cream when the pager went off for an assault victim just down the hill from our house. I immediately knew who was involved; we'd been there and danced with that asshole before. The last time it had taken six of us to hold him down.

I backed the car out of the garage and started slowly down the hill, no lights or siren. Howard and Jeff, the two ambulance lieutenants, were on the way, and I called them on the radio.

"Scott Chief C. to Lieutenant A.," I transmitted.

"Go ahead," came the response.

"I'll see if the P.D. are on scene. If they're not, I'll meet you at the crest of the hill and we can all go in together."

"10-4," Howard answered.

I was glad they were coming. Howard and Jeff had worked together so long they could finish each other's sentences. Howard had dark graying hair and a walrus-like mustache that he could re-grow in about a day and a half after he decided, on occasion, to shave it. Jeff had lighter hair, but the same kind of mustache, and an insatiable appetite for all things NASCAR.

As I reached the driveway of the old farmhouse, I couldn't see a police car, so I continued on toward our rendezvous point. A few yards past the driveway, my headlights caught his shape, our friend from the last time, as he jumped into the roadway in front of the car, trying to flag me down.

Shit, this is not good. If I passed him and kept going, the next car down the hill might be driven by a woman or young kid, and God only knows what this asshole would do to them. Having little choice, I pulled over and got out, careful to take my portable radio with me and, more importantly, my heavy metal five cell Maglite.

I approached him slowly, trying to keep some distance between us.

"What's the problem, Snake?" I asked.

"You gotta help," he stammered. "My uncle's hurt bad."

"We've got help coming. You just calm down."

He repeated his plea for help.

Then like right out of the fucking *Exorcist*, his voice changed into this strange whiney otherworld type sound and he started yelling, "My uncle is the devil," repeating it three or four times. His voice then changed back to his normal—okay, nothing was really normal about this whack job—tone and pitch.

I slowly reached for the microphone on the radio, keyed it, and said, "Lieutenant A., get up here now!" as softly as I could. I knew Howard and Jeff had received the transmission when I heard the deep throaty roar of a car kick into overdrive at the bottom of the hill.

At this point, Snake decided we were going to be friends; the only problem was I didn't want to be his friend. He stepped forward to put his arm around my shoulders. Up until that point I had been careful to keep the end of the Maglite in his chest, maintaining a good distance of separation. His little move changed all that.

Well asshole, time for your teeth to come out. I side-stepped a bit and started around with the heavy metal flashlight. I was going to hit him as hard as I possibly could, smack in the mouth, and regain separation. I don't know if he saw me start to swing at him or just changed his mind, but he stepped back, which saved him substantial dental work.

Howard and Jeff pulled up next to me at that point. "I'll keep him here. You can check the house. Be careful," I warned needlessly. They'd been here before, too. Fresh out of patience with this loony, I grabbed him by the front of his shirt and forced him down into the ditch alongside the road.

"Sit there and don't move!" I told him.

Howard and Jeff pulled past and into the driveway of the farmhouse. The cop finally arrived and stopped next to me. I reached over and grabbed Snake again and yanked him to his feet. He was big enough to snap me in two any time he wanted, but he was as docile as a puppy by this time. I opened the rear door of the police car, and shoved him into the back seat, slamming the door shut. He wasn't going anywhere now.

I walked up to the house to see how Howard and Jeff were doing. They hadn't gone in far, seeing right away that it was a crime scene. We stood in the kitchen staring into the adjacent room; as I had suspected, the uncle was dead. From the look of things—primarily the blood—Snake had bounced the poor old man off every wall in the room. All four walls had red streaks and stains, some up near the ceiling. A large television rested on the old man's head.

We started a log of who was in the house and where they had gone. This would be helpful for the police. The paramedics came in and put the uncle on the monitor to verify what we already knew, and we logged it.

A few minutes later, the county district attorney came in. "Who's the guy in the back of the patrol car?" he asked.

"The guy who did it," I responded.

"I'm liking this more and more," the D.A. said.

We handed him our log, which pleased him to no end, and hung around while the state police evidence technicians photographed and processed the scene. Then we helped the coroner load the old guy up.

The local police chief was on scene by now as well. We were sent back to the station to write out our statements while everything was still fresh in our minds.

The trial a couple of months later was anticlimactic. The judge slept through most of my description of what I saw. There was no real dispute that he had killed the guy; the only argument was whether it was first or third degree murder. The D.A. charged him with first because it was an election year and he didn't want to be seen as soft on elder abuse crime.

The district attorney's prep for the trial was simple. "Tell the truth about what happened, but please don't say he's nuts."

My testimony was a joke. I expected the defense lawyer to use the *Exorcist* voices from the asshole to show diminished capacity. Never came up. He asked me how far away I lived and how long it took me to arrive.

"No further questions, your honor," was his next statement. My jaw hitting my chest followed; I was so surprised.

That outcome, however, wasn't a surprise. The judge took all of about fifteen seconds to find him guilty of third degree murder. The twelve years he got may have surprised Snake, but nobody else.

• • •

It makes me feel old to admit she's a full-grown woman now. I call her Smiley; what I've called her since she was a baby. She is the daughter of Steve, the department chief, and his wife Jennifer, who worked at the dairy bar. The first time I held Smiley, though, came as a surprise.

Late one afternoon I was getting cleaned up after work when the pager went off for a child that stopped breathing at the dairy bar. I knew just about everyone who regularly went there, so this wasn't good.

I jumped into the car, threw the red light on the roof, and flipped the siren as I pulled out of the driveway. My right foot pressed hard on the accelerator, quickly traversing the two miles I had to cover.

I turned into the driveway, tires spinning in the gravel, and stopped amidst a cloud of dust in front of the building. I ran into the restaurant where Robert, one of the owners, had a frantic look on his face.

"In there," he gestured toward the kitchen.

Jennifer held the convulsing baby in her arms. I grabbed the infant and started to open her clothing so I could check her for respirations and pulse. She was breathing but unconscious, her tiny form quivering with the seizure and hot with fever. I pushed through the kitchen to the outdoors where it was cooler. As I got outside, Jeff, one of the ambulance lieutenants, was pulling in. I handed the baby to him, and we jumped in my car. Tires throwing gravel again and siren yelping, I called the ambulance.

"We're en route in my vehicle. Meet us at the ramp to the interstate," I ordered.

We pulled in at the same time as the rig. Jeff carried her to the back and put her on the stretcher where we got oxygen on her and continued to monitor her vitals. The medic unit pulled in, and the paramedics checked her as well. So far so good.

Another vehicle pulled in. It was Steve, her father, the department chief. He climbed in with us along with the other ambulance officer, Howard, who arrived at the same time.

Jeff jumped out to take my car, and Howard went to get Jennifer to drive her to the hospital in Steve's Blazer. Steve and I rode the ambulance with the medics down to the emergency room.

Smiley was fine. She had convulsed due to a high fever. They kept her at the hospital for a day just to make sure, though. When everything had settled down, we could laugh about it, especially remembering how the ambulance officers ended up shuttling cars around while the fire chiefs rode the ambulance.

• • •

Another call in the same area had less happy results. "Report of a house fire, people trapped," the dispatcher said. This middle of the night call was the kind that we dreaded. Heavy fire blowing out of the ranch house's living room picture window showed how much of a head start it had before anyone called. Even worse, we knew there was an old man in there somewhere. I had the attack. Steve helped the new Fleetville chief organize the overall scene and water supply.

With no apparatus on scene yet, I entered the basement through the garage. There was a bedroom down there as well; empty, as a quick search proved. I started up the interior stairs to the main floor to see if I could get in there to reach the guy. At the top, but with no air pack, I could go no farther. The door frame was burning, with fire licking over my head into the stairwell, and heavy flames beyond.

I called out, "Fire department, anybody there?" The crackling of the flames was the only response.

Back outside, I had a good idea of how I wanted to proceed. As the engine pulled into the front yard, I grabbed Nicky who had ridden in the officer's seat.

"Blitz it with the deck gun right through the picture window. We need to try to hold that wall by the hallway and keep it out of the bedrooms," I told him. I had a crew ladder the eave and pull the slatted vent out at the top corner of the attic while another crew stretched a hand line. I knew there was fire in the attic over the bedrooms. We needed to push that back before guys entered to search. Having the ceiling or roof collapse on them would have helped no one. We don't trade firefighters for civilians.

With water flowing into the attic, we took out the rear bedroom window, reportedly the old man's room. Conditions were improving, but still weren't great. Snotty dark grey smoke was pushing hard from the window opening. I couldn't just order somebody in, so I went myself. Howard came through the window behind me.

Searching quickly, we covered the room, Howard holding my boot as we checked different areas. Visibility sucked, there was smoke right to the floor. My hands ran across the sheets of an empty bed. The floor surrounding it was clear also. When we reached the doorway, my hands suddenly struck something soft. There was only one thing that felt like that: a body. It was the old man. I bent closer and shined my light on him. He wasn't burned so much as baked, the oven of a house turning any exposed skin a dark ugly brown. *Damn it, we're too late.* My shoulders sagged. He was gone, likely dead before the call was placed.

What could I have done better, faster, more efficiently? He was lying only about five feet from where I stopped at the top of the interior stairs, just around the corner from them. Five fucking feet, but not knowing he was there. Second guesses flooded through my mind.

Later we found out the family had taken the smoke detectors down to paint the various rooms and they weren't put back up. The fire had started in a detached garage and spread to the house through a breezeway. All but the old man were awakened by the fire. It was so advanced that the rest of the family had to jump from their bedroom windows. It took them a couple of minutes to reach the neighbors' house, wake them up, and call 911.

At one point, the old man had woken up and rolled out of bed. He only made it about four feet before being overcome with smoke. The coroner agreed he was probably dead before the call was placed. It still didn't help much. Time and again I've been over the decisions. Those five feet might as well have been a mile. Professionally, I know they were correct, but being so close and not getting to him still frustrates me to this day.

• CHAPTER TWELVE •

Learning the Country Style

The white-over-red Peter Pirsch engine pulled slowly out of the narrow bay door of the old station, roof-mounted four bulb light rotating, sending out beams of red into the darkness. Nicky had the wheel and I settled into the officer's seat. The tanker nosed out of its door as well, diesel exhaust streaming from the big Mack engine.

Valentino's, the old White Swan restaurant was on fire; the glow in the sky visible for miles. The multi-story landmark was on Route 6 in the valley. It was at the base of the roadway known as the Wildcat. The units on scene had maxed out the available supply from the hydrants. We were responding to set up a tanker shuttle to supplement that supply.

The trip was quick, with no traffic and a downhill run almost all the way. Nicky pulled the engine off to the side and out of the way so we could find out what we would be doing, and where they wanted us to set up. I went looking for Chief B., who had command of the fire.

Finding him in a doorway on the nozzle of an inch and a half hose line was a bit disconcerting, but I tried to remain nonjudgmental. He explained where he wanted us set up on scene and what we would be feeding. No problems so far.

"We can set up the fill site at General Dynamics," I told him. The plant was right around the corner and had a totally separate water supply we could use for the tankers.

"No, I want you to fill at Sugerman's," Chief B. replied, indicating a department store just up the road.

"Bob, that's the same main you're already off," I tried to explain. I knew it would just screw things up if we filled there. He insisted.

What the hell. It's his fire.

Steve went to work setting up the dump site, and an engine was directed to Sugerman's as Chief B. had decided. Tankers began dumping their water into the drop tanks and the first few headed up the road to be refilled.

As the fill site began flowing water to refill the first empty tankers, screams began over the radio. Predictably, pumps on scene were going into vacuum, the water coming down Route 6 being diverted to the tanker fill engine. I sought out Chief B. again. He was unhappy his plan wasn't working out.

"Just let me set it up at General Dynamics. I guarantee I can make it work," I told him.

He acquiesced just to get rid of me, I think. I made arrangements for Steve to send me an engine, and Sonny, the Justus chief, and I took off over to the plant to begin setting things up.

We parked and walked into the lobby security desk, our full bunker gear startling the half-asleep desk officer.

"Hi," I started. "Get me Reggie on the phone, please. I want to use your fire pump." I smiled at the now totally confused guard. He just stared at me until I motioned for him to pick up the phone and begin dialing.

The guard woke him up and I motioned for him to hand me the phone.

"Reggie, it's Gary. I need to use the fire pump."

"Huh?" the confused reply came from the plant fire marshal.

"Reggie, look out the window." I knew where he lived, which wasn't far from the fire scene. I heard him set the receiver down to go to the window. Just as quickly, he picked it back up.

"Holy shit! I'll be right over." I probably didn't need the phone to hear him, his response was so loud.

I called Steve on the radio and told him he could start the engine over. Sonny and I went out and down the side road where the one thousand gallon per minute fire pump and two hundred thousand gallon water tank were situated. With this kind of supply, the engine would be nothing but a big gate valve.

Reggie and the engine arrived at about the same time. Reggie unlocked the pump house while the engine crew hooked up to a hydrant right outside. They laid out four three inch lines, which gave us the capability to fill two tankers at a time. When we opened the hydrant, the fire pump came to life, and we were in business.

A couple of hours and a hundred thousand gallons of water later, we secured the operation and broke down. The plant and corporate people were thrilled with the whole thing. Their equipment had been of service to the community; they even put the story in their internal newsletter. We were happy because we moved a lot of water with very little effort compared to a normal shuttle where we would have to draft from a creek or a pond. Even better, it was the last time Chief B. told me something couldn't be done.

• • •

I was once again in the officer's seat of the engine, this time on a house fire near Chapman Lake. The assistant chief who should have been on scene doing a size-up and setting up our initial attack was down by the lake shore, screwing around with a gate, which would ultimately allow later incoming units to get to the water.

With him out of the picture, I had to size it up on arrival, and saw we had moderate to heavy smoke coming from pretty much every opening in the house. Since smoke doesn't usually like to go down, that gave us at least an idea that we had fire in the basement.

We pulled a line to the cellar door, a half-assed Bilco-type arrangement. The line went hard with water as we prepared to pop the door. Face pieces on, black viscous smoke poured from the now-open entry door. I was behind the nozzleman who started down the short flight of stairs and then stopped dead in his tracks.

"I'm taking smoke," he yelled through his face piece. Whether he was, or just got too scared when visibility went to zero, who knew, but there was no time to discuss his situation.

I took the nozzle as he backed out, and the third man, Rick, became my back up. We advanced into the five foot high crawl space, inching forward on our knees into the blackness. After progressing about ten feet or so, I could see a red glow over to my right, but a bit further in.

I opened the nozzle, and quickly knocked down the flames. Visibility was still nonexistent, but there was no heat or visible fire. Holding the nozzle with one hand, I squatted with my other holding the underside of the floor joist above me. I had Rick yell back to the doorway and tell them to get some ventilation started. We waited. It seemed like ten minutes or more but it was really only about two. There was still no visible fire, no significant heat. There had been no back up line at the door when we went in, so there was no sense going in any deeper at that point. Besides, we assumed that being in an open space, we could see everything there was to see, the smoke notwithstanding.

Whhooossshhh! From nowhere and yet from everywhere, fire rolled over the top of us. My hand dropped to the bail on the nozzle and snapped it wide open, as I fell backwards onto Rick. I started whipping the nozzle around to keep the fire off of us. Turning my head, I could see a problem.

"Rick, help me turn the line. It's getting behind us," I yelled through my face piece. At that point, I knew we had to fight our way out. It's a simple ratio—if the fire gets behind you, the odds of getting killed skyrocket.

We got the hose line turned, and knocked down the flames behind us, as we started pushing for the door. I kept the nozzle moving all around us, and Rick humped the hose behind me. We reached the doorway and crawled up the stairs to the outside.

The fire had extended to the first and second floors, and we chased it around for a while. I was irritated by the lack of a backup line, which could have assisted us in the cellar.

After everything was out, we looked at the crawl space from every conceivable angle to try to figure out the flash fire. No definitive explanation was ever obtained. One theory, which was possible, but not necessarily probable, was carbon buildup on the wood joists from multiple puff-backs from the furnace. While we never did figure out why it had flashed, at least nobody had gotten hurt.

The main thing that saved us was a lesson I had learned from the fire in Maryland where I had been injured years before and gone out the window head first. The nozzle man there had put the "nob" down. While nobody could predict what would have changed if he hadn't, we knew one thing: flowing water is always better than non-flowing water. That lesson—don't put the nozzle down—has stayed with me and in this case, it saved our asses. It's something I drive into my son Mike's head every time we train. Being able to get the nozzle open almost assuredly let us get out of there without getting burned, or worse.

• • •

Most of the arson fires we see are set by adults, but not all of them. In this case, it was a twelve-year-old boy. This kid was being interrogated by Wade, the police chief, a man not only carrying a large gun but towering a good two feet over the youngster. The boy was ninety pounds of solid attitude. Observing this performance, I noted he kept repeating a few weird lines about fire. Then it hit me; the kid was quoting lines from the arsonist in the movie *Backdraft*.

There were two fires in a house on the Justus side of the township inside of an hour. The first one was out on arrival; we never made it to the scene. The second one was out as well, but we did make it in.

Pawing through the scorched paper and trash on the bathroom floor, it didn't take long to find the partial remains of the book of matches used. That's when the fun started with the kid. Originally, the first fire had been written off as accidental. We knew better now. Looking around the house, it was obviously not a stable family situation.

It was one of those Monday holidays that some people get off and others don't. The Pennsylvania State Police Fire Marshals did, and no amount of talking could convince the sergeant at the barracks to authorize the overtime needed to call them out. This kid needed their immediate attention, but he didn't get it. We were frustrated when we left, no one was sure if we'd be responding there in another hour or so. Everyone was convinced of one thing, though; we would be back eventually.

It took two weeks. It was daytime again, when the adults weren't around. From the dispatcher's report it sounded as if the kid had done a much better job getting a fire started this time. I heard Steve call out about the same time I did. I knew he hadn't been on the previous calls and wasn't aware of the entire situation. He would arrive first, so I called him on the radio.

"There are a bunch of kids living there. Make sure they're all out," I told him. He acknowledged, and went on scene with a working fire. Apparently the kid was getting better at his trade.

I pulled in and put my gear on; by the time I reached the front of the building, Steve had finished counting noses, and confirmed all the kids were out and accounted for. All of them, including the babysitter, were upset, save our twelve-year-old arsonist. His face was serene.

For the Justus assistant chief in charge that day, it was his first time in command of a working fire, and the task looked overwhelming. We divided things up, like always, to make it easier. I took the interior; Steve, the exterior; and Nicky, the water supply. With that command structure in place, all the incident commander had to do was supply us with resources; simple, but hard.

I packed up and took two crews—each with inch and three quarter lines—into the house. Steve worked on getting the building laddered, and Nicky worked on setting up the fill and dump sites. If we did our job right inside, we wouldn't need much of Nicky's water, but if we didn't, he'd be a busy boy.

The fire was on the second floor. We started up the stairs, one crew at a time. The first crew hit the fire in the hall, and then went to work on the room on the left. The second crew pushed the fire back further and held the hallway. We were all in about a ten square foot area and could barely get off our bellies, so intense was the heat. The kid had definitely done a hell of a job.

I worked the radio mic up to my face piece and called Steve. "Get the roof open," I yelled through the plastic and into the microphone.

"Working on it," the response came. We heard the saw start up above us. *We should get relief pretty quickly.* The crews were measuring progress in inches at this point.

Guido, a great firefighter from Fleetville, was on the roof and and began opening a big hole. To give us the relief we needed, he had to get the ceiling above us open. Pike pole in hand, he began hammering downward to breach it and allow the heat to escape. It sounded like he was having a bitch of a time. We found out later that the ceiling was actually tongue-and-groove wood, not the typical plaster or dry wall. Guido eventually got an opening, but if it weren't for his incredible strength, he never would've succeeded. Strong doesn't even begin to describe him. More than once, I saw Guido take four cinder blocks, place them in a square, and pick them up with one hand where the four corners met.

Once we had some ventilation, we got the fire knocked down fairly quickly. It looked like the kid had taken it to the next level. There was evidence of a flammable liquid pour on the floor, which explained why the heat was so bad.

The fire marshals were available that day and confirmed that he had used alcohol for his pour. Like many calls, we didn't ever find out for sure how things turned out. One story we heard was they got the kid into counseling. Another was that the family relocated, and he simply became some other fire department's problem. We never went back there, so something definitely happened. I hope the first rumor was right, because he was one scary kid.

• • •

Getting hurt in a fire is far from rare. Even with all the concentration on safety, which can help reduce the injury rate, shit happens and will continue to happen. I've been lucky in that the injuries I've suffered have been of the minor variety. The pain caused by telling my mother or, more recently, my wife, has usually been worse.

On one such occasion, I had taken a vacation day to get some work done around the house, the most pressing of which was taking Mike's crib apart. It was time to move him to a big boy bed. Sitting down in front of the crib with a small assortment of tools, the disassembly had just begun when my pager went off for a house fire on Franklin Valley Road. It was quicker to get to the scene than the station, so I left in that direction. It didn't take long before I could see the column of smoke in the air.

Steve, Guido, Ross, and I all arrived at about the same time. There were plenty of firefighters, but no apparatus had arrived. The house was a single story ranch with a two car garage attached to the house via a breezeway. The garage was fully involved with fire blowing through the roof and spreading into the house through the connecting breezeway.

Ross eyed the roadway between the house and the neighboring Grumman plant, a nearby manufacturing facility. The fire pumps and large suction tank on site fed the sprinklers and yard hydrants for the building. It was just close enough that maybe an engine could lay from one of the yard hydrants to the house. Looking at it quickly myself, I knew one thing: if it worked it would be close. Ross didn't have long to make his decision, the siren from Engine 63 was now audible in the distance.

"Chief 63A to Engine 63," he called over the radio. "Lay in from Grumman." The decision was made, and we now had to hope it was right and there was enough hose on the engine to reach. Ross was a stout shit, typically with a plug of chew behind his lower lip and an empty beer can for his spit cup.

I grabbed the air pack that Ross carried in his truck and slung the straps over my shoulders. Steve had a pack on as well, so now all we needed was an engine; ours was coming from the opposite direction. We started pulling an attack line off Fleetville's engine. Ross had been right, but barely. There was about thirty feet of hose left in the bed—less than a length—just enough to pull off and hook up to the intake on the pump. You can't cut it any closer.

Our engine started gunning the main body of fire in the garage with the deck monitor. Steve and I advanced the inch and three quarter line to the front door. We needed to cut the fire off inside and push it back toward the garage.

With shitty black smoke pouring from the opening, we knelt near the front door and put on our face pieces, air now flowing into the masks. Gloves and helmets now in place, we picked up the line, now rock hard with water, and crawled into the living room. Steve had the nozzle, and we advanced into the blackness. He veered to the right, looking for fire; so far nothing.

"I'll look in further," I yelled to him through my face piece. His "okay" came back equally loud and muffled.

I crawled toward the rear of the house, careful to maintain orientation with a wall while away from the security of the hose line. I didn't have to go far. Entering the kitchen, I had no problem locating the fire. The room was rolling with flames. Retreating back into the living room, I located the line again and gave it a hard yank. Steve crawled back toward me with the nozzle.

"The kitchen's off," I yelled to him as we worked to turn the line in the tight confines of the living room. We crawled into the kitchen, flames rolling over our heads, and Steve opened the nozzle, directing the heavy straight stream at the ceiling. Moving in further, we knocked down the bulk of the fire in the room.

About halfway across the linoleum floor, a large chunk of the dry wall ceiling landed on our helmets and backs. It didn't hurt, but it did get our attention.

Rolling over to get the debris off us, we could see that the now-open ceiling exposed significant fire in the attic space directly above us. Since we were now half on our backs, it was a simple matter for Steve to open up on the attic and quickly knock down the spreading flames.

That finished, I took the nozzle, crawling closer to the breezeway, which had been the conduit of the fire as it spread into the house. Keeping the nozzle moving, we only had a few more feet to go.

The nomex hood I was wearing was old and the elastic around the face opening was shot. It didn't fit snugly, leaving gaps around the face piece of the mask, and should have been replaced. A quick blast of superheated air or steam entered the lower portion where it drooped away from my face piece. Pain shot across the front of my neck burning my exposed skin. It wasn't deep or deadly, but it hurt like a bitch. As quickly as it came, it went, the sharp pain replaced by a steady throbbing. Luckily, another thirty seconds of water and we were done.

Steve and I exited the house to drop our packs. The pain had eased up on my neck and I figured it was probably just a first degree burn—a really bad sunburn. I was more worried about the hell I was going to catch from Michelle when I got home. Jeff was standing by with the ambulance, and when I walked over to him he knew it wasn't a social call.

"What's wrong?" he immediately asked.

"I think I got burned," I told him.

He examined my neck and concurred with my diagnosis.

"Yeah, but it doesn't look too bad."

"No shit. I know it's not bad. Can you do anything for it?" I asked.

He thought a bit. "We've got some coated gauze we're not supposed to carry anymore that should be good."

I didn't ask the obvious question. If Jeff thought it would work, though, I trusted his judgment. He dug around in a compartment in the back and quickly returned with a roll of gauze in a weird looking package.

"It has some sort of coating that helps work on the burn," he said as he wrapped my neck with a couple of layers of the stuff. Damn, that felt good on my scorched skin. I didn't care if we were supposed to use it or not; it sure felt fine to me.

The overhaul was completed while I kept Jeff company in the ambulance; shortly thereafter, we were released. Now it was time for the fun part.

When I got home, Mike was running around playing while Michelle worked on this or that. Everything came to a crashing halt when she saw the gauze. My explanation went a lot faster than her lecture. I think the phrase, "You have a child now," was used at least fifty times.

It could have been worse. I guess Mom could have been visiting that day.

• • •

If there is a cliché for fog thicker than pea soup, we had it that morning. Driving over twenty-five miles an hour was pushing the envelope as we responded to a school bus accident. It was shortly after the scheduled start time for the school day, so we were hoping the bus had no children aboard.

After the excruciating crawl to the scene, twice as long as it would have taken under ordinary conditions, I pulled in to see a van t-boned into the side of a school bus. The good news was the bus had been empty, except for the driver, who was uninjured; the bad news was that the van's driver wasn't.

Climbing into the passenger side of the van, I found the driver pinned by the crushed dashboard. He was in agony. "Please help me," he moaned.

Young Todd crawled in from the rear, and we began clearing the plastic console pieces away as best we could. Once we got closer, I held his body up while Todd ripped away the pieces of dash crushing his lower torso.

"Stay with me, pal. We're gonna get you out," I told him.

"Don't let me die," he begged.

"Just hold onto me," I said.

We were going as fast as possible. Another piece or two had to be moved. We almost had him. I looked over at his eyes, and just watched him drain away. He died in my arms.

Ten seconds later, he was free. Todd and I pulled his body from the driver's seat as carefully as possible, laid him on the floor in the back of the van, and started CPR. We had to try.

The paramedics climbed in and took over, got him in a long board, and transported. Todd and I sat on the van's back bumper, drained, drenched in sweat, not speaking. There was nothing to say. We had done everything we could, but it wasn't enough. Sometimes, that's just the way it is. The internal injuries were too severe.

The van driver was at fault, the cops told us. Driving way too fast for the foggy conditions—over fifty miles an hour, the cops said—he didn't see the bus until it was too late. I'm sure this was no consolation to the four or five young kids he left behind.

There is an interesting dichotomy following a bad call. If a brother firefighter asks casually, "Are you okay?"—the required response is, "Yeah, I'm fine." It's simply a way of letting each other know we care. A firefighter understands and isn't looking for a different answer. However, if a civilian asks the same question, it pisses me off; it shouldn't, but it does. Civilians haven't seen, smelled, or touched what I have. Somehow, within my twisted psyche, that means they have no right to ask the question much less to expect an answer, especially at that moment. And I certainly don't want to answer any questions from a reporter looking to plumb the depths of my emotions for the titillation of their readers or viewers. When it's happened, which isn't often, I just mumble something incomprehensible and walk away.

• • •

One fine Saturday morning, we were working on an addition to Steve's house. We all worked on each other's home improvement projects over the years. It was free labor supervised by someone who actually knew what they were doing. I was part of the free labor portion of the crew.

We were putting up board insulation for a cathedral ceiling and not having fun fastening it in place. Each missed hammering of a nail, which was easy to do when hammering upside down at an angle on the underside of the roof, created a hole in the insulation. Misses were as common as hits, and our frustration was building.

Bang—miss. *Damn.*

Bang—miss. *Shit.*

Bang—miss. *Fuck.*

Our pagers all went off simultaneously, a cacophony of shrill beeps in the confined room, the dispatcher announcing a house fire in Fleetville. Tool belts clattered to the floor where they lay abandoned as we headed for our vehicles, deciding who would ride with whom.

Only a couple miles from the scene, we arrived to find an old two story farmhouse with heavy fire blowing from the windows on the first floor. It probably was on the second as well, but we couldn't tell at this point. We put on our gear while awaiting the first engine.

Once the engine pulled up, I put on an air pack and grabbed the nozzle of the big three inch line and, with Artie behind me, started up the steps for the front door. Artie was a big bull of a guy, incredibly strong, which was a very good thing. There was no way could I handle a line like this without someone like him behind me—that bad boy could flow upwards of five hundred gallons per minute. With him there, it was like having a garden hose in my hands...sort of.

We advanced through the door and into the front room where we were greeted by a parlor full of fire. I opened the nozzle and with the huge volume of water we were flowing, it was like hitting a candle with the forementioned garden hose. Okay, maybe not quite that easy, but it was still incredibly fast fire control. Continuing our advance, we killed more fire on the first floor until our bells went off.

We exited the building and pulled our helmets and face pieces off our sweat-soaked heads. Additional crews had other lines now being stretched into the house. I walked over to the engine where air bottles were being changed. Dropping down to my knees and leaning forward, I provided access to the bottle while I rested. Thirty seconds later, a new cylinder was in place; I was, in theory, ready to go.

I walked over to the Fleetville chief running the fire to find out what needed doing at that point. The good news was I could rest a few more minutes. The better news was I would partner up with Guido and work the second floor.

We grabbed some hand tools and went in. The first floor was knocked; guys were working on opening up. We located the stairs and up we went, finding a line lying there from a previous crew. It looked like they had done a good job. The small amount of fire had been knocked down.

We started looking around for places the fire might be hiding—the walls, the ceiling, etc—and began poking holes. As we found it, we'd open up further and knock it down with the line. The rooms were surprisingly clear, visibility was excellent, and the work was easy. It looked like we were dealing with just small amounts of fire which didn't seem to be running fast.

Guido was standing on a bed in the main bedroom where the fire had been, opening up the ceiling above him, when we heard multiple air horn blasts outside. This was the signal to evacuate the building, typically reserved for a serious situation. We looked at each other and shrugged. Not a flicker of flame was evident to us, but it is a cardinal rule never to ignore an evacuation order. Somebody, we figured, could see something we couldn't.

That said, the conditions imparted no sense of urgency, so we picked up our hand tools and strolled down the stairs to the first floor. We glanced around; the first floor crews were gone, so we were the last two in the building.

Walking out onto the front porch, we could see the chief practically jumping up and down, gesturing toward the roof, and firefighters running around moving lines. Still not knowing what the hell was happening, we walked out into the front yard and removed our helmets and face pieces, and turned to look at the house.

Understanding rapidly set in. Fire blew through the roof, a column of flames thirty feet in the air. It had found its way past us into the attic at a point away from the room we were working. Perhaps a pipe chase or opening in the exterior wall was the culprit. The chief, rightfully so, was worried about the roof remnants collapsing into the second floor—where we were.

The exterior lines made quick work of the fire, allowing fresh teams to reenter and finish off the attic from below. Then the hard work of picking up all the hose and equipment began.

Hours later, we returned to Steve's. The tool belts were still lying where they'd been dropped. No one had the energy to put them back on. A hot shower and a cold beer—that's all we wanted.

• CHAPTER THIRTEEN •

Kids Are the Worst

It was a late afternoon in the spring of 1997. Michelle was out running errands and I was at home with the kids. At the time, Mike was five and Megan was three. Late afternoon is a bad time for most volunteer departments and we were no exception. Most first shift workers aren't home yet and second shifters have left to go to work, making manpower an issue at that time of day.

My pager suddenly went off for a structure fire on Shields Road. The kids ran for shoes and coats and I helped them finish dressing and get into the car. Down the hill we went and over onto Shields Road. I looked for the address and eventually found it; it was a trailer down a fairly long driveway, lower than the main road. I got out of the car and put my gear on, and called the county on the radio to assume command and give a brief size-up.

"Lackawanna, Scott Chief A. on scene with heavy smoke showing, establishing command," I told the county over the radio as I watched the dark gray smoke belch from the trailer.

One of the fire police showed up at this point. Instead of directing traffic as he normally would, I had a slightly different job in mind for him.

"Smackie, would you stay here for a few minutes and watch the kids for me?" He smiled and nodded. It wasn't the first time over the years he'd had to do that for somebody.

I walked down the driveway to the trailer. It was obviously not being used as a residence, but I wasn't sure what it was being used for. I called the engine on the radio and told them to lay a supply line down to the trailer. I told the rescue to pick up the lay and set up a drop site. I had the tanker come down with the engine for an initial water supply of over four thousand gallons until the rest of the tankers on the first alarm arrived. Fleetville was sent to the pond at the farm down the road to set up a fill site. Most of this was precautionary; if things went the way they should, we wouldn't even use half of the water from the first engine and tanker.

The engine started down the hill, the five inch wide yellow hose peeling off behind it. It pulled sideways and stopped, the crew dismounted pulling an inch and three quarter attack line. I told them which door to take and in they went once they had water.

Additional manpower started to arrive including Ross from Fleetville and a backup line was pulled. Ross worked nearby and was able to leave for serious fires. The first line didn't seem to be moving much; a basic rule is if the line isn't moving forward, the fire's winning. It was frustrating to watch.

"Ross, take both crews, get in there, and put this fucking thing out," I told him. I knew the interior would be in good hands with him.

I continued to watch and listen to some of the radio traffic while the fill site was being set up. With Ross now inside, conditions were improving. The smoke turned white—telling me we had water on the fire—and shortly thereafter, it began to lift and then clear. Ross came out and told me it was knocked down. Now I had to revert to my other, more important role—being a Dad.

I walked up to the road and the kids were still watching from the car, wide-eyed, under Smackie's supervision. Conditions were safe enough that I could take them down with me and they could stand next to the engine and watch. One on each side, we held hands and walked down. They stood perfectly still next to the engine while I dealt with the details of the fire. Soon we were able to start returning companies and began breaking down the fill and dump sites.

Both kids were polar opposites at that age. Mike was shy and didn't talk much while Megan, like her mother, would talk to a dead man. Actually, not much has changed since then. After a while, Billy, the pump operator, took a break and bent over to talk to the kids. I didn't hear any of the conversation but when he stood back up he was laughing. Later I learned he had asked how they were doing. Mike was silent, knowing his little sister would talk for both of them. She proudly announced in her three-year-old voice, "This is my first structure fire." It wasn't the last.

• • •

Some kids didn't just watch fires. Another spring day, a group of teenagers had taken a day off from school. Unfortunately, they didn't go hang out at the mall or do whatever the hell normal kids who skip a day of classes do; these kids decided to torch a few things. The barn looked like the only building we had a chance of saving. There were four or five cars fully involved; a couple of other smaller buildings burning and on the verge of collapse. Brush was blazing all around. There was stuff on fire a couple hundred feet along the dirt road in both directions.

The torches had a field day at this place. We didn't have time to worry about it, though.

Fleetville's engine was on scene, and as I walked up I could see a line off of it snaking into the barn. About this time, one of our members, Mick, came barreling out of the partly open sliding barn door. I grabbed him.

"What's going on?" I asked.

"It's bad in there," he said. Mick was physically a big guy and a bit of a bullshit artist. If you listened to his stories, he was the best firefighter in the company—but not after today.

"Where's your partner?"

"Steve's still on the line," he said.

Now I was pissed. You don't ever leave your partner, ever. Not bothering to grab a pack, I followed the line in to make sure Steve was okay. He was, but he didn't have a pack on, either. He had the line almost to the high bay section of the barn. It was rocking in there, a solid wall of flames visible through the doorway. I backed him up, and we got the line into the opening. Steve had the nozzle on straight stream and kept it moving. It was so hot he just kept his head ducked down; from behind him I could see a little bit better.

"Up, left, over, right, down," I said as he'd move the tip in response to my directions.

We started to gain control of it, and although it seemed to take forever, it was only a couple of minutes; luckily the smoke conditions weren't as bad at floor level. Once the high section was knocked down, we took the line to the loft and hit the small amount of fire that had extended through the barn board wall. It was out, at least as far as we could see.

Back outside, other firefighters were finishing off the cars, brush, and outbuildings. Steve and I were trying to figure out what had gotten into Mick. It was warm, but the conditions were certainly bearable. We were left with one conclusion—we could never trust him again. When somebody does something like that once, you never know if or when they will do it again.

Grey smoke continued to seep from under the metal roof of the barn. We looked inside and saw no smoke or fire, but the underside of the roof was wood, which meant there must be a void between the two. We were nervous about putting someone directly on the roof based on the damage we could see to the underside. After talking about it a bit, we were able to convince the chief to call for the ladder out of Clarks Summit. It was the closest stick and small enough to fit in the limited space we had. We were hoping we could work off the stick to pull up the metal roofing material and get to the fire underneath.

The aerial arrived and was positioned in front of the barn. It was a weekday, so their crew was light. We also didn't want to ask them to do the dirty work for us. Steve and I were the only two who had worked off an aerial before, so we were elected.

We both went up the ladder, but found it had a couple mechanical issues we hadn't anticipated; for example, there was a problem with the hydraulics. The driver was actually adding hydraulic oil as we tried to work. That didn't exactly give us a warm, fuzzy feeling. Similarly, the ladder wouldn't move with both of us on the stick at the same time. We found it was sluggish but okay with just one of us. We would have to take turns.

I climbed down the aerial to the turntable while Steve went first. Slowly, the two of us, alternating on the stick, ripped up the metal roofing. It was nailed to furring strips, which created the void the fire was burning in, as small as it was. The installer had not been stingy with nails, the panels secured on about six inch centers. It was ass-busting work.

Finally, we put it out. Mick was nowhere to be found, but we weren't upset he was gone. There were no formal repercussions, but word spread about his actions. I never went in with him again. It didn't break my heart, if it kept me alive.

• • •

The motor home blazed away in one of the southbound lanes of the interstate, the column of black smoke visible for miles. Traffic was at a standstill, the area above the fire resembling a parking lot. We had to access the scene by going northbound in the southbound lane. Even with the right safeguards and assurances that traffic is shut down, the hair the back of your neck inevitably stands on end as you go the wrong way. All it takes is one car to squeak through to ruin your day.

The fire attack on the Winnebago seemed to be going smoothly. The engine and the fire in the driver's compartment were knocked down pretty quickly, but there was some lingering fire underneath.

I laid down on the blacktop to get a look beneath the vehicle. The problem became instantly clear. The gas tank was leaking and the dripping fuel pooled beneath the camper was burning briskly. The solution was obvious as well, and I hollered over my shoulder to get a foam line set up. In the meantime, the crew kept the plain water line moving to sweep the burning fuel from the underside of the frame.

The foam line reached the side of the camper. Opening it up near the side of the vehicle until the foam flow was established,

I began to slowly move it in toward the fire, creating a blanket of the white stuff as it rolled onto the burning gasoline. The foam began to smother the pool fire, but the section in the center where the tank leaking was not going out. Bouncing the foam off the tank itself did nothing to help.

I yelled back to Steve to arrange more foam. For fun, I'd shut down the line and watch the fire flare back across the blanket then open the line again, pushing it back to the spot being fed by the leak. We weren't going to be able to put the fire out while the tank was still leaking, so this actually preserved a little of the foam for us. Howard was lying there beside me and he wasn't enjoying the game I was playing with the fire.

"Knock that shit off," he said, nudging me in the shoulder.

Steve decided he'd better get the propane tank out of the camper in case we couldn't hold the fire or, worse, ran out of foam. He leaned in and began working, his head buried inside the compartment of the camper. A minute went by, then another. When he finally reappeared he yelled, "It won't come out."

"Shit," I replied. "How much foam do we have left?"

"Not sure, but I'll check," he said. Steve was the most mechanically talented individual I knew. If he said it wouldn't come out, it wouldn't come out.

Howard was still with me and getting a tad worried. He knew that without the foam, the pool fire would expand again, eventually impinging on the immovable propane tank.

"What do we do when we run out of foam?" he asked.

"I don't know about you, but I'm going to be about a quarter mile down the road if this thing blows," I responded, but we kept at it.

Pretty soon Steve tapped me on the shoulder saying, "We're down to the last pail."

Sometimes it's better to be lucky than good. The flow from the tank slowed to a few drips, then stopped. Within seconds, the foam blanket took over and covered the remainder of the spill until it was out. We could breathe again. Fortunately, we didn't have to jog down the highway.

• • •

Living in a small town, we regularly deal with people we know on emergency calls. I was working my way through a pile of paperwork in my home office when the pager went off for a trailer fire up the dirt road from our house. Running upstairs to grab my keys, I glanced out the window expecting to see a column of smoke, but there was nothing in sight.

In the car, the tires spun as I punched the accelerator, speeding up the long driveway leading to the trailer. Black smoke was belching from the rear door. The homeowner was in the yard, frantic as she watched the fire.

Pulling my gear on, I grabbed the five pound dry chemical extinguisher from my trunk. As I approached the door, I could see that the bathroom was rolling with flames licking into the hallway. Crawling in to stay below the flames at ceiling level, I pulled the pin on the extinguisher and aimed for the center of the room. The fire began to darken as I whipped the tip of the black rubber hose toward the ceiling and back toward the floor as the bottle emptied. I couldn't see any more fire but the smoke was as thick as snot and tasted like crap. Reaching in, I found the door and was able to get it closed, knowing if there was still fire in there, as I suspected, I had at least bought us a few minutes.

Retreating to the yard to get a breath of air, I saw another one of our guys pull in and start to put on his gear. I grabbed the radio microphone and called the engine. They weren't even on the road yet because they didn't have a driver.

Shit. The ambulance was coming back from another call with most of the day crew on it, including the few guys who could drive the engine. The only good news was that Engine 24 was on the road. After a minute or so, we could hear their siren screaming as they came down Country Club Road.

Keying the mic again, I told them to lay the driveway. Jerry, the other guy who'd shown up, and I tried to get a garden hose going off the washer bib inside, but the fire had already tripped the breaker for the well pump so there was no pressure. The smoke kept pumping, so we knew the fire in the bathroom was growing again.

We heard Engine 24 start up the driveway. "If they have a crew, we're golden," I told Jerry. "But if not, be prepared to do this the old fashioned way. I'm not losing this fucking place now after all this work."

The old Mack came over the rise, hose peeling from the rear bed. My heart sank. Only the driver was visible. The engine pulled to a stop and we grabbed a pre-connect, stretching it out and flaking as we went. The line wiggled like a snake as it went hard with water. Jackie, one of Greenfield's guys, ran over to us in street clothes. He had no gear with him. I had no idea how he'd gotten there, but I didn't care at this point.

"Can I do anything?" he asked.

"Get the window on the other side," I told him.

He looked around and saw a shovel lying on the ground and grabbed it. I knew the window would be taken care of.

Jerry and I advanced the line to the now closed door. *Crash!* The window and door went almost simultaneously. He opened the nozzle at the ceiling. Thirty seconds of water and we were done. It wasn't pretty but it was out.

The ambulance was on scene now with the rest of our crew, the engine at the bottom of the hill pumping Engine 24's supply line. Old Billy and Jackie had both been at the bottom of the driveway along with a large neighborhood dog. When they started up the driveway the dog started chasing them. Both of them figured they didn't have to beat the dog, just the other guy. We were lucky we didn't have two cardiac patients by the time they got to the top. Neither of them knew that the dog wouldn't hurt a flea.

I knew Michelle would be getting home about then and I also knew I needed to suck some oxygen. I didn't want to walk down and do it at the ambulance where she could see me, so Sharon went down and got a portable oxygen rig. It didn't matter. Michelle figured out who it was for when she saw it being carried up. I heard about it the rest of the night and on numerous occasions since.

The trailer's owner was very upset. Billy, with his usual flair for public relations, told her, "Lady, don't be upset. We never save trailers." After she had calmed down, she was very happy with the results. The trailer would need some repairs, but it would be fixable and livable in short order. She embarrassed the hell out of me the next time I saw her, throwing her arms around me and giving me a big hug while exclaiming, "My hero!" It was just nice that someone appreciated what we had been able to do.

• • •

Hugs and thank yous don't happen often. This is a thankless job. So why do it? Why be a firefighter? The reasons obviously differ for each person. For me, it's the satisfaction of doing something only a small percentage of people can do—entering buildings being consumed by fire and having the skill to save lives. You sure as hell don't do it for the gratitude. If you start to do this thinking people will regularly thank you, disappointment is in your future. I can count the number of times I've been sincerely thanked on both hands with fingers left over. In more populated areas, the victims you see are mainly strangers. That doesn't make the job any less difficult, only different. In a small town, you regularly see past "victims." Most would rather forget about the incident; some may not even recognize you from the semi-anonymous bunker gear, and others, I think, simply don't know what to say.

•   •   •

While there is a personal satisfaction that is derived from most calls, others are devoid of anything positive or uplifting. The worst calls, for me at least, involve kids. One sunny summer Saturday morning turned into a nightmare when we arrived at a two car accident. The mother's piercing screams were like a knife twisting in my guts. She dangled from the driver's seat of the overturned minivan, held in place by her seat belt. The van rested on its side.

"I killed my baby!" she screamed over and over. Beneath the passenger side door, the torso of a little girl extended out into the roadway, her head still under the vehicle.

This was not the time for air bags or fancy jacks. Four or five of us leaned into the side of the van and lifted. We slid the little girl from beneath the vehicle, but went no further. There was nothing we could do. Her head had been crushed like a grape by the van as she rolled out the passenger side window. She hadn't been wearing her seat belt. We removed the mother, with minor physical injuries, and the ambulance transported her to the hospital. Everybody was upset, myself included. But you can't—and don't—show it at the scene. It's just not done.

After the coroner pronounced the little girl dead, Guido and I placed her in a body bag. I went home and sat on the front porch and drank a bottle of wine. Dumb, but that's what I did. The alcohol didn't do anything for the memory, but it gave me something to do while trying to digest things. I still don't like to think about that call. I can't even imagine what the mother must have felt like.

• • •

There was another little girl I'll never forget. The coroner had pronounced her dead and enough photos had been taken to fill an artist's photo exhibit. Nobody but the most ghoulish would want to attend this one, though.

She had been located a good hour into the fire. The initial search team had been driven back by the intense heat. They had gone in as far as they could; possibly too far, passing a couple of rooms of fire with nothing but a pressurized extinguisher, two and a half gallons of water in their hand.

When I got there, the search crew was inside and a line being stretched to the top of the exterior staircase. There was fire blowing out the top of the doorway, bright red flames licking skyward. This was not good.

"Push that fucking line in there now!" I screamed at the crew on the stairs. "There's two guys in there." They opened up, but were not pushing in and I was not quiet in making my frustration with their lack of aggressiveness known. At that moment, the two guys from the search crew emerged from the doorway and were yanked onto the exterior porch, their turnout's smoking. They hadn't found the little girl.

The attack crew advanced a little ways into the second floor and then retreated, reporting that the floor was soft. I took another crew into the first floor, and we killed a small amount of fire there. Structurally, the underside looked okay, but we had to work with the information and reports we had.

The building was constructed in two separate sections, a front and rear. The bulk of the fire was confined to the rear section. We knew the front was fine. It's a difficult decision, but after a time, it becomes clear that based on the fire and smoke conditions, there is no longer anyone to save in the building. It is important to make sure guys aren't taking dumb risks to save the unsavable.

We set up an operation with two lines over a ladder through the picture window on the second floor. We established good accountability and air management and started to work from the good section toward the rear. After three crews, the bulk of the fire was knocked down and the third group located her small form lying in the hallway leading from the bedrooms.

Her father, who had made it out, had told her to follow him. I will never understand how he could have done that. Why not put the child in front of him where he could see or maintain control of her? It doesn't matter though, because it doesn't change the end result. A beautiful little nine-year-old is gone.

• • •

We had a number of arson fires in the Deer Lake area of Fleetville over the span of a couple of years. The M.O., as the cops would say, was the same for all of them. A pour of flammable liquid was made on the outside of a cottage, luckily always unoccupied, and then lit. The Pennsylvania State Police Fire Marshal had a suspect, but wasn't able to prove anything. There were no witnesses, and she—yes, the suspect was a woman—wouldn't confess.

One night, the fires got the attention of a larger audience. As usual, one of the unoccupied cottages was well involved, and we had exposure problems with the fire threatening the adjacent homes.

Getting there was an interesting experience. The roads were twisty and narrow, and there was no light on this moonless night, save that of the flames in the distance. As I drove into the area, I could see tire tracks leaving the main road in line with the flames I could see in the distance, and assumed this was the route to the fire. Following the tracks, I found myself in the center of someone's front lawn—not exactly where I wanted to be. I backed out and continued on the main road, eventually finding the fire.

After donning my gear, I found Steve and explained the unusual route I had taken. He looked a bit chagrined.

"Those were my tracks," he said. "I ended up in the same front yard."

The power lines were creating difficulty getting to the fire. They were an exposure themselves, and a danger to the firefighters. Ross called the Comm Center via radio and had them request the power company to kill that section of the grid. I was impressed; that technique was new to me. In a matter of two or three minutes, all the lights went out in the surrounding cottages. It would've taken a minimum of a half hour for the "trouble truck" from the utility to arrive and disconnect the specific line giving us the problem. I figured I could add that to my toolbox for the next time I ran into this situation.

After that, the fire was routine and knocked down fairly quickly, albeit with significant damage to the torched cottage. The next day, however, I heard enough to remove that new tool from the box.

The utility was, in fact, able to kill the power remotely, and from where we were standing, it looked like a quick and localized operation. Quick it was; localized it was not. Ross started getting complaint phone calls early the next morning. His little radio transmission had resulted in a power outage well in excess of one hundred square miles, a bit more than he had bargained for. It was explained to him in no uncertain terms by the utility and the county—both of which were also fielding complaints—

that in the future, only someone's imminent demise was the necessary criterion for pulling the trigger on such an order. Otherwise, based on the number and tenor of the phone calls he received, his demise would be imminent.

• • •

The old bar called Mumbles was located right at Fleetville Corners. Around four in the morning, I pulled in about the same time as Engine 63, having been dispatched a few minutes earlier for a report of the building on fire. After slinging on air packs, Guido and I stretched a line to the front door, and he gave it a couple of kicks with his big right foot, but it didn't move. I stepped back over to the engine and asked Pop, the pump operator, for the irons, a Halligan bar/axe combination. He told me they were already off the piece, I asked him for an axe—also not available. He handed me a pike pole. Figuring it was better than nothing, I walked back to the front door, and wound the thing like a javelin, striking the door right above the lock. Amazingly, on the third blow, it let go, and we were in.

We entered the front room and saw some fire behind the bar. Guido popped it, and we made quick work of all the flames that were visible. We started around the bar to check out the glow we could see back there, but something about the heat and smoke conditions made me want to slow down. I started pulling the ceiling with the pike pole I still carried. Sure enough, there was a lot of fire up there and it was running. We killed it through the opening I had made, and moved in and opened up some more. Guido knocked that down as well. We worked our way across the entire dining area of the bar like that. We didn't want to leave fire behind us, particularly over our heads, as there was no sign of a backup line yet.

Ross joined us behind the bar, and we could see the hallway back into the kitchen rolling with flames. Guido's bell started ringing, so he handed me the nozzle and bailed out. I knocked down the fire in the hallway leading to the kitchen. All we needed to do was make the turn and we'd have it.

Ross grabbed my coat and told me to hold up. "They're taking the outside door to the kitchen and are gonna hit it from there," he yelled through his face piece. He knew we didn't want to have opposing lines, so we squatted in the hallway. The wait was interminable. He called for an update over his portable. They were still working on the door.

Frustrated, I yelled to him, "C'mon, ten feet and we have this knocked." He knew it, and I knew he knew it. But orders were orders, and in this case, freelancing could get somebody hurt. We continued to wait until both our bells began sounding and it became a moot point. By the time they got the door open, the fire had extended to the floor above.

Out on the street, we began getting our bottles changed. More apparatus and manpower were on scene now. By the time we had fresh cylinders, they had finally finished off the kitchen.

Guido and I took our line back in and found the stairs from the bar to the second floor. There was solid fire at the top, which we killed as we advanced up the stairs. At the top, there was fire to both our right and left. Guido knocked down what we could see on both sides, and started in to his right. I grabbed his coat.

"Where the fuck are you going," I yelled through my face piece. He motioned to the right. "No way," I yelled. "You don't know how much fire is left on the other side, and we don't have a second line here yet." He understood, and we worked what we could see from the top of the stairs until another line came in behind us and we could move in.

After our second bottles we sat on the bumper of one of the rigs watching the action. There had to be at least a hundred guys there by then. Ross kept repeating over and over, "Awful lot of clean goddamn people here." He was pissed that we were going to go back in and do some overhaul when most of the crews seemed to be just watching.

We were all a lot younger then. Three cylinders today would kill me. *Those were the good old days.*

• CHAPTER FOURTEEN •

Not Your Call

In 1999, I took over as chief of the Scott Township Hose Company after serving many years as assistant chief. It was an interesting time with some young guys coming up who wanted to learn. I had the chance to mold both them and the department. The chief's job is different from any other job in the department—you are responsible for everything, every single call, whether you're there or not.

One Sunday morning my pager went off while I was perusing the newspaper and relaxing. The dispatcher announced the units for a full structure fire box before reporting a house explosion. The location was only a couple of miles away so it didn't take me long to get there.

"Lackawanna, Chief 36 on scene with a working fire. Two story single family dwelling," I reported over the radio.

I could see flames in the living room. After giving the initial report, I ran up the front steps. The porch door had been blown off its hinges, so I yanked it out of the way. The fire in the living room was growing and spreading rapidly.

The engine arrived on scene and the crew stretched a line to the front door and started to don their face pieces. The fire was now on the porch blowing out the door. I stood on the front lawn with the engine company, continuing to watch the fire develop. They were ready. I slapped the nozzle man on the shoulder.

"Go kick its ass for me," I told him, looking him hard in the eyes. I wanted him to know I was totally confident in him. He nodded and picked up the nozzle and started up the front steps. He opened the nob on straight stream and pushed in on the fire aggressively, whipping the tip around, getting good conversion as the crew moved in.

I called Engine 24 on the radio. "Pull a second line to the front door on arrival."

We got the water supply started, the rescue laying down from the main road to Chapman Lake for a tanker fill site.

The bulk of the fire was knocked down quickly. The first crew did a great job. It took a lot more work to overhaul, though, as the fire remained in a number of nooks and crannies.

Once it was out, we were able to take a closer look for the cause. We knew there had been an explosion due to the damage to the walls. They were blown right off the foundation. In the crawl space, we discovered the culprit—somebody had taken out the furnace. The whole house was being heated by a space heater on the first floor, which was fed off the same gas line that had once gone to the furnace. When they cut the line to the furnace they didn't cap it; they simply crimped the copper with a pair of pliers. The crimp leaked and the crawl space filled with gas that slowly worked its way up. When it made it to the first floor and found the flame on the space heater—BOOM—and we were off to the races.

Good thing the guy who owned the house was drunk when it blew him off the couch. He might've gotten hurt if he was sober.

• • •

Much of the work on the fire apparatus was done in house, although we tried to do a better job than the highly skilled plumber from the house explosion. An early winter night right after supper, there were four or five of us working on upgrades to the brush truck. It was a great time of year to do this work as brush fires were few and far between. I wanted to get a true two and a half inch discharge off of the pump to take full advantage of what it was designed to do. We were working in a garage right around the corner from the station, which had a lift suitable for the access we needed. Mike was about eight-years-old and he came with me the same way I used to tag along with my dad.

Shortly after we got started, everyone's pager went off for a chimney fire assist to our sister company over the hill. I had no real intention of going as in the early evening I assumed we'd have plenty of people, but figured I'd take a run over to the station to make sure the rigs went out properly staffed.

For whatever reason, when we got to the station there were few firefighters and no officers at all, save me. Young Todd was going to drive the engine, and he was a very safe driver, so I decided that Mike was going for his first ride on a fire truck. I picked him up and tossed him into the front seat of the engine then pulled my bunker pants up and swung on my coat. Climbing in next to my wide-eyed little boy, I used an air pack strap to secure him in place, and nodded to Todd to go ahead.

Out the door we went, siren blaring over the silence of the winter night, red lights shooting beams into the early evening darkness. Todd took his time, partly because it didn't sound like a serious call, and partly because Mike was riding between us. Mike's little eyes were wide open taking everything in, fascinated as I yanked on the air horn cord when we approached an intersection. He didn't say a word the entire ride.

When we pulled into the scene, I lifted him out of the cab, and swung him up to stand by the top mounted pump panel next to Todd. Our sister company had two pieces on scene already, so there was no need for us to pull any lines yet. It didn't take long to determine that we wouldn't be needed. The neighboring chief released us, and Mike went back into his air pack strap.

With the time we had lost on the call, we didn't get much work done on the brush truck. When we got home, all Mike wanted to do was tell his mother about his trip on the fire truck. I thought I'd be in some hot water with Michelle, but she was okay with it. She knew I would never let anything happen to him.

A day or so later, I ran into Todd. He was still amused by the whole thing.

"When you tossed him into the front seat it was like right out of the opening scene from the movie *Backdraft*." Thank God it didn't end the way that fire did.

• • •

Mike wasn't the only one who went to fires; late one afternoon it was Megan's turn. Michelle and Mike were out together, and

Megan was home with me. The pager opened for a trailer fire in Fleetville. Down Route 438 we went, under the interstate and up the hill past the state park. As we neared the intersection with Route 407, we could see the column of black smoke in the air. Megan was excited; she was going to see a good fire.

Approaching the trailer park, it became evident that the trailer was well involved in the first row off the entrance road. So instead of going into the park and putting another vehicle in the way, we parked at the little general store across the street from the complex. Megan, about five-years-old at the time, watched the fire excitedly while I put my gear on. She grabbed my hand as we crossed the road together and I found a nice safe spot for her near some other kids who were watching the fire. I told her not to move an inch until I came back for her.

Fleetville's engine arrived, and an initial line was stretched. From my perspective, the involved trailer looked pretty well gone, and our main purpose was to knock it down to protect the exposures. Siding and windows were already starting to melt and crack on the neighboring units and the situation was made even more difficult by the fact that Fleetville arrived with only three guys.

Leo was on the pump, and his father, Lenny, had the line in his hands. He typically functioned in a fire police capacity so it had been a while since he'd been on a line. He tried though, hitting the trailer from the end, attempting to darken it down and lessen the radiant heat. I jumped on the line with him. We weren't moving, which reduced the progress on the fire quite a bit. I tapped him on the shoulder and asked if he would like me to take the nozzle for a bit. Most guys will never give up the nozzle, but since he wasn't thrilled to be there in the first place, he happily acquiesced.

We switched places, and I told him, "Just grab onto my coat and stay with me." I opened the nozzle and started down the side of the trailer. We were able to get up close and drive the stream right into the main body of the fire, darkening it down substantially. Unfortunately, that also increased the smoke, and since Murphy's Law works everywhere, that's when the wind decided to shift. We were halfway down the side and suddenly inundated with smoke.

Lenny hung right with me, which was good, because we could no longer see each other. Clean air was available at either end of the trailer. It wasn't going to take any more effort to go forward and finish it than to go back, so forward we went. We killed more fire until we reached the end. Naturally, the wind shifted back again and our whole side was nice and clear. In any case, the bulk of the fire was knocked and we had no more exposure problems.

Lenny was happy as a clam. It had been years since he had made an attack on a line. It brought back some good memories for him. I went to find Megan and take a break. I wanted to make sure she wasn't worried, since all she would've seen was her father disappearing into the smoke and flames. I didn't need to worry. She was more concerned about whether we would make it home in time for her basketball practice than she was about me.

I was sitting near a supply line and she was standing next to me. Ross came over to move the line and, not recognizing Megan said, "You need to stand over there, little girl," motioning to an area across the road. "We've got to move this hose."

"It's okay, Ross," I said, "She's mine."

His jaw dropped. "Oh my God, the last time I saw her she was this big," he said holding his hand about two feet off the ground. "She's beautiful. You're in trouble when she gets older."

As a proud father, naturally that made my day; even more than getting the nozzle on the fire.

• • •

I tried to be careful what calls the kids went to as I didn't want to expose them to the ugly side of things at a tender age. One wreck initially sounded routine. Michelle was out, and I had the kids, so I just sat back to listen. But the call didn't stay routine for long. The lieutenant got on scene and started screaming over the radio. The air was crowded with too many people trying to talk. When he finally got some clear air, it became obvious why he was stressed—there was a van into a car with fire and entrapment.

I couldn't just sit and listen, but I also had to be careful what the kids saw.

"Get your coats on. We're going for a little ride," I told them.

We took the fastest slow ride we could. I waited until I got close to the scene before turning on the red light to get past the stopped row of cars.

I parked and got the kids out, and we walked up to the side of the rescue, away from the wreck, all holding hands. Steve was up at the pump panel with the headset on, running things. He had been delayed in getting there as well, but not as much as I had. I climbed up so he could fill me in on the details.

The victims had already been transported, the one who had been pinned in the vehicle on fire was in critical condition with horrible burns. The crew on the rescue had knocked the fire down, set up the tools, extricated him, and had him in the back of the ambulance for transport in eleven minutes. Based on what I could see of the vehicles that was awesome work and I was extremely proud of them. Based on the fire conditions, the amount of damage to the vehicle, and how badly the victim had been trapped—nobody could've done it faster—nobody. Making it even more impressive was the relative youth of the crew; they were far from our most experienced guys, but had been training hard, and it paid off.

I assumed command so I could stay near the kids and Steve took care of making sure the fire was out and the tools and equipment were loaded back on the apparatus. Once he was finished, I wanted to get the kids home and then get down to the station to talk more with the guys. There was no rush; the roadway would be shut down for hours until the police finished conducting their investigation. The Pennsylvania State Police accident reconstruction team had been requested by the local cops.

Shortly after I got the kids home, Michelle returned, and I explained what was going on as delicately as possible. She got the kids ready for bed, and I left for the station.

The crew from the rescue was there, trying to decompress a little the way we normally do, by talking about it. Not about feelings or emotions but how things went on the call itself. You relive every second and go over every minute detail; not necessarily to second guess, although that happens, but to get the stress out. I'm not sure why it works, but it does.

One of the guys was second guessing himself in this case. The victim had third degree burns over about eighty percent of his body. If he lived, he would be horribly disfigured and probably lose some limbs. "Maybe we shouldn't have…" he started. I stopped him right there.

"We don't get to make that decision," I told him. "It's not your call. That's for somebody else to decide, whatever your beliefs are. We go out and do the best we possibly can and let the chips fall where they may."

A stress counselor came in and talked to the crew for a bit. The police called her in to give a group session for the responders. They're not used real often, and nobody had to stay, but most everyone did. I thought it was overkill, but if it made them feel better, what the hell. I had heard the screams of someone burning before, but the young guys hadn't. After she was done talking and listening, though, I had my say.

"This was a bad call," I began. "But how many times do you read in the paper about an almost identical incident someplace else? It was bad, but not uncommon and you need to remember, it could happen again right here in the next fifteen minutes. If it does, we go out the door and handle it again the same way." It was kind of a "suck it up" speech. I didn't believe in babying them. I saw a few heads nod in understanding.

"You guys did a hell of a job out there and you should be proud of yourselves 'cause I'm proud of you."

The victim did live, though he was horribly disfigured and crippled. He actually visited the station once afterward to say thanks, and the topic came up again of maybe we shouldn't have. My answer was the same.

"Not your call."

After you become a father, things change a little bit. Every time you go out the door, there is the thought in the back of your mind that you might not come home. It's not something you think about consciously when the pager goes off; hell, if you did, you'd never go on a call. It is present in your subconscious, though. Once you take on the responsibilities of a father, coming home becomes even more important. It doesn't really affect how you do things on the fire ground, but sometimes it makes you think a second or two longer before making a decision, which is not a bad thing.

• • •

The citizens we protect sometimes make bad decisions which necessitate our involvement, like working underneath a car without using jack stands. One time the address was right around the corner from my house. A man was pinned under a car. I started down the hill and made the right onto the main road. I found the number on the mailbox and pulled in the driveway. Scully, the township cop, was right behind me. We both trotted up to the house where we could see a car that looked like it was being worked on, lying beneath it was a man. The frantic neighbor who placed the 911 call was there. She knew enough to give the jack a pump or two just before we got there, which was just enough to get the car off his chest. Scully and I dragged the victim out from beneath the car. His face was black, and there was no pulse or respirations. I ripped open his shirt and began CPR compressions as Scully started mouth-to-mouth. We worked him for a couple of minutes and his color started to improve. We heard a slight cough and stopped CPR to check him. *Yes, he was back!* He wasn't ready to go dancing, but he had a pulse and shallow respirations.

We turned him over to the medics and our ambulance crew, and a few weeks later he walked out of the hospital. There were residual effects, but at least he was alive and functioning. Successful resuscitations are few and far between. This one felt good. It was only the second time I worked on someone in cardiac arrest who ultimately survived.

• • •

Good calls make the time and effort worth while, especially when they come as a surprise. The older I get, the more I stereotype college kids in various ways, and then get reminded that they aren't always true. The call was for a dorm fire at nearby Keystone College, just over the county line. The fire was drawing more and more resources from throughout the area. We were dispatched for our cascade system to fill air bottles. Arriving on campus, I threaded my way through a throng of students and parked on the lawn above the still burning dorm.

The handsome four story building's pale red brick walls were marred only by the smoke and fire currently blowing out of the window of one of the rooms on the first floor. I made sure the guys on the cascade knew where they were to report and, satisfied they could take care of business without me, looked for a company to hook up with. Clark Summit's aerial was setting up in the lower parking lot, so I walked over to see who was on it.

Joel, the assistant chief, was on the scene along with three other guys.

"You need any help, Joel? I'd rather work with somebody I know, and it looks like you're about it here," I said.

"Hell yes," he responded. "That'll give us two crews of two."

He got orders to send manpower to the roof to open up and start a secondary search inside. I was assigned to the roof team, and while Joel got the stick up and positioned, we gathered our tools. The direction we received was not to cut the roof, just get any existing scuttles, skylights, etc. opened up for ventilation.

My partner and I started climbing. As we got closer to the top, I could see this wasn't going to be fun. Ladder 4 was a short seventy-five foot stick. The end fly of the ladder didn't even have sides. It was at full extension, and there was only about one rung over the roof line, not the four or five I would have preferred. There wasn't going to be any "hold onto the ladder and step onto the roof" type maneuver available to us.

I reached the top, and tossed the irons I was carrying five or six feet onto the roof. Then I half dove, half rolled off the ladder onto the surface. My partner did the same. I looked back at the ladder unhappily. At the time, it was our only way back to the ground. Getting back on that bitch was not going to be fun. We gathered up our tools, and started looking around for vent opportunities, tapping the surface as we went looking for soft spots. With the fire three floors beneath us, we weren't expecting to find any, but it was better to be safe than sorry.

The only significant opening we could find was a rooftop bulkhead door, which we quickly forced open, popping the lock with the Halligan bar.

"I hope they get that fire out downstairs, 'cause that's how I'm going down," I said, pointing at the stairs inside the door we had forced. "I'm not getting back on that fucking thing." I gestured back toward the barely visible ladder tip at the side of the building. My partner laughed and agreed.

We hung around on the roof for a bit, listening to the radio transmissions below us. Eventually we were told to come down and, as I had planned, the stairs were the route of choice. Reaching the first floor, we took a look at the dorm room that had been off. The room was gutted, and the fire had burned through the top of the door and out into the hallway. It had been starting to get into the room above via the window, but the company on the second floor had done a good job checking this extension. The bulk of the manpower had been used to search the three floors of dorm rooms. We were already starting to hear that the college management was pissed about the number of broken doors and locks. *Tough shit.* If one of their kids had been taking an afternoon siesta, or snuck back to their room for a quickie with the girl-friend and been trapped by the fire, I don't think a few broken doors would have bothered them too much.

The kids looked at it a bit differently. The campus had been the scene of a few other recent fires in unoccupied support build-ings. Although it wasn't being advertised, these kids weren't stu-pid. They knew the same as we did that there was a torch running around. They also knew that we had been there for them in the building they lived in.

I was walking up the hill back to my car with a couple of other guys who needed a ride, my air pack slung over my back, helmet tipped back, sipping from a bottle of water. We were sweaty, dirty, and smelly; not exactly movie star material. We were passing through groups of kids hanging out on the hillside, who had been watching the firefighters work in their dorm. It started slowly, but as we walked through, a few kids began to clap, then a few more, then the rest. This wasn't some smart-ass response to authority figures; it was real. They actually appreciated what we had done for them. Shivers and tingles ran up and down my spine. It was fucking incredible. I felt ten feet tall.

• • •

I remember the same feeling on another call around the same time. There was an old man who liked to take the same walk every day; through the farm field he once tended up to the old stone wall, which marked the field's edge. Generations before, farmers gathered the large flat stones exposed by their plows, piling them along the edge of the field, creating, over time, walls delineating the property lines. This particular farmer's daily routine had gone on for some time, but was getting more difficult, as the old man suffered from both Alzheimer's and Parkinson's.

Shortly after four p.m. on a brisk spring day, my pager went off with a report of a missing elderly man. I hoped we'd find him quickly; not only to reunite him with his worried family, but because Mike had a Little League game in an hour that I wanted to attend. I arrived at the farmhouse around the same time as the cop on duty. We spoke briefly with the frantic wife and daughter, and learned some of Mr. B.'s history. He had left around lunchtime for what would ordinarily be an hour's walk, and had not returned. Before calling us, the family had searched his normal route themselves, to no avail. The amount of time that had passed since he was last seen was a concern. I pulled the cop aside for a brief conversation.

"Look," I said, "this could be considered a police matter, so if you want to run this thing, just say the word, but tell me now."

A look of fear came over his face. "No, you guys should handle it," he quickly responded, happy to avoid the responsibility. *So much for the Little League game.*

That settled it, and I started to organize things. "Lackawanna, Command 36. Give me manpower from Stations 28, 24, and 18 for assistance, and Field Comm 20 to the scene." Over the next ten to fifteen minutes, apparatus and manpower began to arrive. We started to establish teams and divide up the area to be searched. I wanted no freelancing, and also no areas missed. In the farmhouse, the family had an aerial photo of the farm, apparently taken by one of those commercial services that sell it to you through the mail. It gave us a great initial perspective of what we were looking at. I sent the cop to the township office building for the tax maps of the area, which would also show us some different things as well as the neighboring properties.

"Break in if you have to," I told him, "but get me that fucking map." He disappeared and, within about ten minutes, was back with it. The first companies had been given their assignments and were beginning to search. I called the dispatcher again and requested the local TV station's helicopter to assist in the search. A few minutes later the dispatcher reported that the chopper wouldn't come; they claimed it was needed for the news. *What the hell did the morons think this was?*

Steve, my assistant chief, had a pilot's license. "You want to try this fixed wing?" he asked.

"Can you get a plane?" I asked.

"No problem," he replied.

"Get it. We'll figure out how we pay for it later."

He started for Seaman's Airport in Fleetville with his lights and siren going as he called them on his cell phone, explaining what we needed. When he pulled into the small airfield, about fifteen miles from where we were, the plane was fueled and the propeller turning, awaiting his arrival.

I grabbed my lieutenant, and told him he was going to do staging and accountability for me. An aggressive kid, he always wanted to be where the action was, which today was in the fields.

Every time we sent out a new crew with an assignment, he would say to me, "I'll take them out."

"That's not going to happen 'cause I need you right here," I explained. "Look, Peter, we've got people with each crew that are perfectly capable of managing four or five searchers. I need you to help me keep this organized, and to learn how this is done." At that point he seemed to understand.

We were trying everything in the playbook. I put a call in for search dog teams. There were two teams locally, and both responded. When they arrived, it was obvious they were a bit competitive. I had no time or energy for that bullshit. I explained to them that I really didn't care who found this guy as long as somebody did, so get out there and get the job done. The dogs were given clothing that the old man had worn to get a scent, and shortly they were in the fields working.

Steve was soon over us in the plane, my captain with him talking to me via his portable radio. It allowed us to cover the huge expanses of open fields far more quickly than with search teams on foot or even with quads. It also enabled us to react to and quickly check rumors, which were coming into the command post, of possible sightings of the old man.

I put two divers in the pond next to the barn. We were running out of daylight, and I was running out of ideas. Obviously, if he was in the pond, it was a recovery operation and since the manpower was available and I didn't want to put them in the water after dark, we did it then.

Close to two hours had passed since we started searching. That meant the old man had been missing for upwards of six hours. We started to brainstorm and formulate plans for what we were going to do once darkness fell. We were thinking about trying to get a helicopter from the State Police or a medevac bird, with a thermal imaging camera on board, hoping to locate him via his body heat signature. It was going to be cold that evening and

I didn't have much hope of him surviving the night out there. There was already a light chill in the air, but I could feel rivulets of sweat running under my shirt. It wasn't from physical exertion—I was just sitting in the command vehicle with all of the radios—but from the mental gymnastics and stress of running the incident.

We had crews spreading into neighboring properties and into the woods beyond, much farther than we expected him to go. I was getting frustrated. *What am I missing? Do I have everything organized correctly? Is there something else we should be doing?* We continued to wait.

Just as we were again reviewing our plans, an animated call came over the radio from one of the search crews.

"We see him!" they called excitedly. We silenced all the other teams that wanted to assist, until we could get a good report. "He's conscious, but can't walk." I sent two teams with a brush truck, quads, and a Stokes basket to bring him out.

I stepped out of the Field Comm unit. It felt like a huge weight had been lifted off my shoulders. Backs were being slapped and ear-to-ear smiles were on everyone's faces at the command post. I walked into the farmhouse. The old man's daughter was holding her mother, their residual tears obvious. They looked at me, their faces filled with a mixture of dread and hope.

"We've got him, and he's alive," I told them. Their tears started again, but now from joy.

"Thank you! Oh, thank you," the daughter said, wiping her eyes. Her mother was speechless, trembling with relief. Being able to convey that simple news and see their reaction brought a rush through my system.

The search team that brought him into the farm yard was as happy as I was. Their laughter was contagious, especially after they related their initial conversation with him. He had been sitting on some rocks in the woods away from the tree line, which explained why our plane couldn't see him. When they reached him, the first thing he told the crew was, "That goddamn plane up there is lost. It's been flying around here for hours."

• CHAPTER FIFTEEN •

You Keep It

As the number of injured people in an incident multiplies, so does the complexity of the operation— it could be written as a mathematical formula.

Add the interstate highway as a factor, and the challenges go up. Two nasty incidents in about a six month period reinforced this for me.

The first one was during the day. I arrived on scene to find a minivan overturned in the median strip. The van was on its roof with the occupants scattered like bowling pins around the vehicle. I saw Jacob, a fire captain I knew from down the line, working on a child, doing CPR. He had been driving by and discovered the wreck. Seeing the small, limp body lying in the weeds with Jake's hands pumping up and down on his chest, my heart sank. I had work of my own to do, though. Counting heads, I got a rough idea of the number of injured.

"Lackawanna, Chief 36," I called on the radio.

"Chief 36," the dispatcher acknowledged.

"On scene with a vehicle overturned in the median, establishing command. Give me the next five BLS units on the box and an additional medic unit."

The dispatcher acknowledged, and the tones began their audible dance over the radio alerting the additional basic life support ambulances and paramedics.

Our ambulance arrived on scene, and Old Billy got out and came up to my truck where I had initially established a command post. As tempting as it was to jump in and work on some of the injured, that wasn't my job and giving in to temptation could actually make things worse; it would be like the general running to the front and picking up a rifle instead of managing the battle. Sure, he might take down some of the enemy's troops, but without proper management, he could lose the whole outfit, or in our case, more victims.

"Billy, take triage. I got five more transport units coming. I think that should cover it. Let me know if you need any more ALS."

"Okay," he said as he started down the slight embankment to the vehicle.

Triage is one of the toughest jobs on a call like this; decisions have to be made on who can be saved and who can't or, in more dramatic terms, who will live and who will die. Sometimes it's obvious, but that doesn't make it easy. Other times it's less clear cut and the decision is searing. The EMS personnel have a process to help, but the bottom line remains that these are instinctual gut-wrenching decisions made with the help of training and experience.

I called the county again. "Have all the units report to the southbound side of the incident. All drivers are to remain with their rigs."

That would get everyone facing the right direction to head for the various hospitals. Insisting the drivers stay with their vehicles helped move rigs out as soon as they were loaded.

More help and the additional ambulances I requested started arriving, and things started falling into place. The kid in full arrest went first, and Billy managed the order for the remaining victims. Ambulances were being split up among the three hospitals in Scranton; the most seriously injured routed to the Trauma Center at Community Medical Center. Finally, the last known victim was sent. Our rig remained on scene; it was a shell, having been pretty much stripped of equipment.

We were a bit worried, looking at the overturned minivan. The roof had a lump, about the size of a child's body. It was visible through the smashed windows of the van. We hadn't been able to truly confirm the number of victims, as the family was Indian and spoke little English. The lump hadn't been a priority. We weren't sure anyone was under there, and if they were, they were certainly dead. There were plenty of live but seriously injured people to take care of before we worried about a body.

The wrecker got in position, and everyone held their breath as he rolled the van. It was a clump of dirt. We could breathe again.

• • •

The next incident happened in the middle of the night and it was colder than a bitch outside. Arriving on scene, I recognized a guy I knew from Clarks Summit coming up over the bank. He had been driving by and stopped at the wreck.

"How many do you have down there?" I asked him.

"I'm not sure," he answered.

"It doesn't have to be exact."

"At least eight," he responded.

I picked up the microphone on the radio. "Lackawanna, Chief 36," I called.

"Chief 36."

"On scene with a multi-casualty accident in the median. Give me the next eight BLS on the box and two additional ALS units," I ordered.

The dispatcher acknowledged me. "Chief 36 on scene with a multi casualty incident, requesting eight BLS and two ALS," he repeated to ensure he had it right.

The tones started over the radio for all the units now due. I knew I had plenty of time to get dressed, as it would take a couple of minutes to get them all over the air before the dispatch announcement. We started refining the number of injured. He was right on the count.

We set up operations in a similar way as the fatal wreck earlier in the year, keeping the drivers in their vehicles. I set up my command post in the front seat of the rescue. I could see the wreck and at least it was warm in there.

With the number of rigs that were lined up, it was like an ambulance convention. Ambulances from three counties were parked nose to tail along the highway, red and white lights dancing through the darkness. Portable lights from the rescue focused on the accident scene itself, turning the moonless night into high noon on that small section of the median.

One by one, they loaded and left. We tried to split them up so one hospital wasn't overwhelmed with all of the victims. Finally,

only our rig was left again, stripped to the bones of equipment.
Better news on this one: no fatalities.

• • •

While wrecks present one set of challenges, fires seem to
have them in innumerable quantities. This one was an ordinary
looking house, other than the fact that it was on fire. The tanker
had been the only apparatus requested, and on these middle-of-
the-night runs we would sometimes just send one or two guys
needed to operate the tanker and the rest of the officers would
go back to bed.

For some reason that night I decided to take a ride up. The
tanker crew was experienced so they didn't require any supervision; regardless, I responded, parked, and put my gear on.

Still just a spectator, not having any responsibilities at this
point, I saw the first line stretched into the building by Troy, and
his crew, and other routine tasks were underway. The guys on
the tanker did their thing.

Wallace, the chief running the fire, and Troy's father, saw me
and walked over. We chatted for a minute or so. Then he asked,
"Would you do me a favor and take a crew in?"

"Sure."

I grabbed an air pack off their engine and donned it while
Wally assembled a crew for me.

The house was divided into two sections: a one story side and
another about a story and a half. He wanted us to take a line into
the upper portion of the second section, a routine assignment
until we walked in the front door. I glanced to the side as we entered the enclosed front porch; it was full of compressed gas
cylinders. A closer look revealed they were filled with hydrogen,
not exactly an inert gas. There was enough here for this house
to become the next Hindenburg.

Shit, what the hell are we getting into?

I had a team of three with me. We paused on the porch to make sure the tanks were off. There was a manifold attached to a number of them, with copper lines extending into the building. There was no way I wanted hydrogen leaking into a burning building.

After ensuring that all the tanks were off, we advanced our line into the front room. The amount of heat made me suspect we had fire in the space above the ceiling. One of the guys found a scuttle hole near the stairs that opened into the space. Opening it confirmed there was fire above us, but we couldn't hit much of it from this small hole.

I called Wallace on the radio and told him we needed to get this fire before going to the upper level. I didn't want to leave fire behind us, a good way to get in trouble. Wally agreed, and we started trying to get the ceiling open.

There was a company on the roof venting, which would help. We had a bitch of a time, as the ceiling was tongue-and-groove wood, and it didn't yield easily to our modest collection of hand tools. We were starting to make a bit of progress, though, our efforts resulting in a few small openings in the wood ceiling, when all hell broke loose.

The thermal column dumped like the opening of a furnace door, and flames came down on us through the small holes we had opened. The crew dropped to the floor. Steam, smoke, and heat followed us down, far worse than the small amount of fire. If it weren't for our protective clothing, we all would've known what a lobster felt like. As it was, it was close. We were lucky more fire hadn't dropped down on top of us.

"Everybody out," I yelled through my face piece, counting the guys as they crawled by me toward the door. I followed the last man out.

I already knew what happened. I was hot, and not from the fire and the steam. Exiting the building and kneeling to remove my helmet and face piece, steam started rising from the pores of my bunker gear. It was then that I was met by Ken, one of Greenfield's assistant chiefs.

"Now calm down and don't kill anyone," Ken began.

"All I want to know is who put the fucking line in the vent hole," I told him without screaming, even though I wanted to.

"Just stay calm," he repeated.

"I am calm, Ken, but I'm not going to be if you don't tell me who put the fucking line in the fucking vent hole."

Seeing that I wasn't about to kill anyone, at least imminently, he began to explain. The group on the roof, with Ted in charge, had done a nice job opening things up. Fire was blowing out of the hole in the roof, buying us the time we needed inside to open up from underneath and put it out in the concealed space. Unfortunately, sometimes chiefs get nervous seeing all that fire with no water going on it, so even though it violated all proper practices, Chief Wallace ordered the roof crew to operate their protective hand line in the vent hole. Ken made sure he explained that Ted had questioned the validity and correctness of the order twice before finally following it, and turning us into lobsters.

By this point, Wally had realized his error, and was contrite when I confronted him. It's difficult to maintain the desire to kill when a person is apologizing up and down. As for Ted, it became a running joke, at least on my part, on how he had tried to kill me.

Wally had another shock when he found out about the hydrogen cylinders. The house was some sort of fancy glass lab with hydrogen-based, special atmosphere ovens. One leak while that initial crew was inside would've incinerated all of them and possibly put the house into low orbit.

All chiefs live with the fear that a firefighter or company under their command could be killed following orders they've given, so Wallace was shaken. Especially knowing his son was on the first line inside.

● ● ●

A couple of miles west of the interstate on a dirt road lived a woman known as the Goat Lady. Her house was a menagerie of animals. In its day, the old farmhouse with the classic wrap-around front porch was a handsome home. The Goat Lady had turned it into a barn. The beautiful old front porch, now covered in bedding hay, was where many of the animals lived. Her favorites actually lived inside with her.

It was single-digit bitter cold, the morning we were called. Manpower was short, and the fire was so far along that there was little we could do at that point to save the house. Once the main body of fire had been knocked down from the exterior, we divided the house up to check various sections for structural stability, fire extension, and the like, each chief taking a separate area. I was assigned the rear of the first floor.

I went in the back door, still amazed at the living conditions. There was even hay on the floor in the rear hallway. I continued in a few more feet, noting a couple areas where some residual fire needed to be overhauled. Advancing further, I noted the floor felt spongy, our description for what might be structurally unsound. I flexed my legs a bit to get a better feel for how bad it was. Yep, spongy all right. Then I looked down. The floor wasn't the problem. The spongy feeling resulted from the dead goats I was standing on. A brief shudder ran through me, and then I got back to work. We brought in crews with pike poles, axes, and hose lines and began the ass-busting process of overhaul.

Back at the station, the garden hose came out as everyone took turns hosing off their boots. *Dead goats; there's nothing like life in the country.*

● ● ●

Ted, the Greenfield chief, and I were talking over this and that on the phone the next morning. Ted was a tall, thin guy with a brush style haircut. He was a very aggressive and intense firefighter, and we'd been friends and worked well together for years. Just before we hung up, he mentioned that he had to run down to the city for a bit, and joked that I would be in charge. It was freezing out, bitterly cold; we'd both spent the day before at the Goat Lady's house, and my gear was still wet, all of which he knew.

"Fuck you," I responded. He laughed and we hung up; I went back to my paperwork, and he ran whatever errands he had scheduled. When the pager went off an hour later, I couldn't believe it. The dispatcher announced a structure fire on Greenfield Road.

"Chief 36 responding," I called out on the radio, starting the car toward Greenfield. I already knew who the first arriving command officer was going to be—and it wasn't going to be Ted. Turning onto Greenfield Road from Route 247, the column of smoke was high in the air. I pulled in the long driveway of the farm to see a three car garage with a second story, fully involved. That was bad enough, but the garages were attached to the barn, which was attached to another long one story building.

Shit. I picked up the mic. "Lackawanna, Chief 36 on scene with a working fire. I'll be establishing Command 24. Give me a second alarm."

I pulled up the driveway and parked out of the way. Putting on my wet bunker gear, I began eyeing the driveway in terms of distance. There wasn't a lot of room in the farm yard to turn tankers around if I brought them in here. I heard Ken coming with Engine 24. I had to make up my mind pretty quickly. The engine paused as it turned into the driveway.

"Engine 24 from command, lay the driveway and kick it," I ordered. I wanted the big five inch supply line kicked to the side so we could still use the driveway to get more apparatus to the scene. The engine proceeded, hose peeling out of the rear bed; the crew pushing the heavy yellow line to the side of the driveway.

I held my breath, hoping I had called it right. Was the driveway less than the thousand feet of hose on the back of the engine or were we going to be short? As the engine pulled to a stop, there was one length of hose left in the bed. Close, but we had made it.

"Command 24 to Engine 24, blitz it," I ordered over the radio. I started getting units assigned to a fill site at Heart Lake and the dump site at the end of the driveway where Ken had started his lay. There was a pond near the farmhouse, but it was up hill, through snow and ice, and there was no guarantee we could get anything to it.

Meanwhile, Ken had the gun going, darkening down the main body of fire with the heavy stream, trying to keep it from extending to the barn. More apparatus arrived. Engine 36 added its tank to the mix, blitzing with the pre-connected monitor as well. We started getting a supply line set up between the two engines, as water started down the five inch from the dump site. I had a company stretch a hand line into the barn to cut off any fire extension there. It was going to be a critical operation.

Now that things seemed to be coming together, I walked over to talk to Ken.

"Are you ready to assume command?" I asked him. Ken was still busy pumping their engine.

"No, that's okay, you keep it," he replied. I shrugged my acquiescence, and went back to work.

Engine 18 was on scene by now and Troy looked at the pond by the farmhouse. In addition to the snow and ice, it was blocked by a car.

"I think I can make it if we can get that moved," he told me, nodding at the vehicle in the way. It was worth a try. If he could get their engine in there, it would give us a lot more water to work with.

"Go ahead," I told him. The owner quickly moved the car while Troy climbed into the cab of the engine. He backed the engine up a bit, then floored it. It looked like the start of a drag race, rear wheels spinning and big engine screaming. Then the wheels caught, and before we knew it the engine was on top of the bank at the pond. Troy got out and came over to me while his guys cut the ice to get the hard suction line into the water.

"Uh, I didn't know it was that close," he said. "Another five feet, and the piece would've been in the water," he told me with a big grin on his face.

The line inside the barn was making good progress. My main problem now was frostbite. I couldn't use my radio with my gloves on, and it was so cold that I could no longer feel the tips of my fingers, which were turning a milky white. I kept blowing on them, which helped a little.

I checked in with Old Billy, running the pump on Engine 36. His water supply was good. As an aside, he said to me, "You know, this isn't actually Greenfield Township. The township line follows that old stone wall over to the west. This is actually Carbondale Township." Billy knew more minutiae like that than any person alive, so I had no doubt he was correct. I looked around. Carbondale Township's engine was there, assigned on the second alarm. One of their assistant chiefs was pumping it. I took a walk over and introduced myself. The departments didn't work together often, and consequently we had never met.

"I owe you an apology," I told him. "I just found out this is actually your fire. I'll be happy to go over the various assignments if you want to assume command."

"No, that's okay, you keep it," he replied. It was getting to be a familiar refrain.

The fire in the barn was now out. It had gotten into the loft, but the crew had pushed it back and extinguished it. The main body of fire in the garage knocked down, we were able to go to smaller hand lines to overhaul and kill the hot spots. Lots of work still remained.

About that time, the chief from Carbondale Township arrived. Finally, the rightful owner of the call was on scene. I explained the circumstances and how I had come to have command.

"I was about an hour away when it came in and listened to the whole thing," he told me. "It sounded like a hell of a job."

I thanked him and made my now regular offer. "You pretty well know what's going on now if you want to assume command."

"No, that's okay, you keep it."

Another hour passed and we were picking up and starting to release companies when Ted pulled in with a grin a mile wide. At that moment, we both remembered our phone conversation from hours before. It was all he could do not to burst out laughing. I stood there looking at him still soaked and half frozen.

"Don't even get out of the fucking car unless you're going to take this fire," I told him.

"No, that's okay, you keep it."

• • •

The cold wasn't always a bad thing. It was the Friday night before Christmas, a crisp starlit evening. We were cruising the township roadways with Santa Claus on the rescue. It was an annual event, much enjoyed by many of the smaller members of the community and, truth be told, by many of the bigger ones as well.

The lights were flashing, the siren screaming, the air horn blasting and regular bursts of "Ho, Ho, Ho" were echoing in the night air from behind me. I rode the officer's seat in the cab, just enjoying the atmosphere and the smiling children we encountered on our slow tour. My fun was broken by a radio call.

"Comm Center to Chief 36," the radio query came. After I responded, the dispatcher asked, "You wouldn't happen to be out with Santa Claus by chance, would you, Chief?"

"Affirmative," I answered.

"Can you call in by phone?" the dispatcher asked.

I didn't have a good feeling as I reached for the cell phone mounted on the dash. Was some Scrooge upset by the siren noise, I wondered. When I got the dispatcher on the line, it was nothing like that.

"Hey, Chief, we just had a call from a grandma on Greenfield Road. She was upset 'cause she had been out when you went by and her grandchildren just missed Santa."

"Please tell me she didn't call in on 911?" I asked the dispatcher, almost dreading his response. The 911 emergency line is certainly not the proper method to obtain a visit by Santa Claus.

"Oh yeah, she did," he said with a laugh.

"Sorry about that, we'll take another run down that road."

We have to take care of a grandma like that, I thought to myself.

"Thanks, Chief, and Merry Christmas."

• CHAPTER SIXTEEN •

Take the RIT

After a few years as chief, I decided it was time to give it up; the kids were getting older and were involved in many activities and my work as a fire protection engineer took me out of town quite a bit. Something had to give and it couldn't be the kids or my job. Done correctly, the chief's position takes at least forty hours a week divided up over nights and weekends; I didn't have that time anymore, but I also didn't want to shortchange the job. I felt letting someone else step up was the correct choice. I took a break from an officer's job for a year or so and then came back as an assistant chief for a few more years. The time commitment was far less and there were no administrative requirements and headaches. I really didn't give a damn what color helmet I wore, chief or firefighter.

A number of guys I worked with stopped going inside when they turned forty, some even a bit earlier. I reached that point but still felt physically capable of interior firefighting, albeit at a reduced number of air cylinders. I knew that when the day came, and it would, when I was not capable, I would stop completely. I wouldn't just drive or direct traffic—I couldn't think of anything more boring. I took it year by year; the fires, however, still kept coming.

At around two thirty a.m., I woke up to the urgent beeps of the pager sitting in its charger on the windowsill. The dispatcher's voice began, sending us and Station 24 to the Oscar Black gas station on Route 106 for a reported working fire. I quickly dressed and headed to the car. Mike had just turned thirteen-years-old and had started tagging along to some fires. It was a school night and he had slept through the tones, so I didn't wake him. *Just as well.*

The ride north to Greenfield was eerily quiet, with almost no traffic on the road. I knew I wasn't far behind Engine 24; their ambulance went through Finch Hill Corners as I approached it. I turned left onto Route 106 and headed toward the county line. As I got closer, I started to pick up the radio transmissions from the scene and I heard command request the third alarm. The second had been requested earlier.

Rounding the last curve before the county line, I could see flames blowing from the front of the building. Red and white lights from the first arriving apparatus bounced off the adjacent buildings in the fire-lit brightness of the night. I parked alongside the narrow roadway, well back from the scene so as not to obstruct access. I opened the rear hatch of my Explorer and pulled out my bunker pants while kicking off my shoes. I stepped into my boots and pulled up the pants, tightening the suspenders over my shoulders. I tossed my glasses into the back of the car and pulled my nomex hood over my head. Grabbing my coat and helmet, I closed the hatch and started down the road. I put on my helmet and swung on my coat as I briskly walked toward the flame-lit scene of the heavily burning gas station.

Engine 18 was on the right, or the D side of the main building, which was well involved with fire through the roof and blowing from the windows. I walked past them toward Engine 24. The B, or left, side looked to be something we could possibly save. A line was in place to try to cut off the fire in the narrow alley between the buildings. I reached the engine and threw my accountability tag to the pump operator, hoping the operation to cut off the fire would succeed.

I walked to the B side of the exposure building and saw that smoke was already banking from the eaves, telling me it was too late. Fire had already spread to the exposure. I walked back to the engine and grabbed an axe. Christopher, one of Greenfield's guys, dragged an inch and three quarter line over to the side. We were going to try to push the fire back. He stood well back from the window. Based on the color and behavior of the smoke, we weren't sure if the building would blow or flash when I took out the windows. Carefully standing to the upwind side of the windows, I raised the axe and smashed both panes. The building didn't go but we could hear the distinct sound of fire crackling from the interior.

Christopher stuck the nozzle through the window and we began to apply water to the fire. We could hear it crackling, but we couldn't see the flames through the dense black smoke in the building. While we attempted to knock down the fire, another team began to cut through the main overhead door. An upside down V was made with a saw and they quickly moved a heavy two and a half inch line into the opening. Just as we were starting to see some progress on the fire, the water ran out; not unusual in a rural environment. Everyone backed up and waited for the water supply to be restored. After a few minutes we had water again, but the fire had progressed substantially and was through the roof. The building was a write-off.

I took a break and went to talk to Ted from Greenfield to see what he needed.

"Would you take the RIT for me?" he asked. The rapid intervention team on the incident was a rescue company from Carbondale.

"Sure, just get me a portable," I said and he immediately handed one to me. I went over and met my crew. They seemed like good guys, and they had the right equipment with them.

Due to the volume of fire, an exterior attack was continued for some time. I walked around to get the lay of the land and familiarize myself with the overall fire ground picture. After some of the heavy fire on the second floor was knocked down, the decision was made to try to get a look in the window and see if the owner, who might still be in the building, was in the bedroom. A firefighter searched the area from a ground ladder, but didn't see anyone.

The defensive exterior attack continued for a bit longer until the bulk of the fire appeared knocked down. At this point, Troy, the Clifford assistant chief, came looking for me and suggested that he and I make an interior evaluation of the structural stability prior to committing personnel to complete extinguishment and overhaul. It seemed like a good idea to me, so we donned our face pieces and started up the stairwell on side D. The smoke was moderate; a lot of stuff continued to smolder. Heat was light as the building was well ventilated with significant portions of the roof gone by this point. Troy and I threaded our way through the studs in the wall at the top of the stairs and advanced into the bedroom. Later on, we found out that this was a very good decision as there was little floor left at the typical access point around the corner. We did a thorough search of the bedroom and confirmed that the owner was not there.

At that point, we decided to take a look at the underside of the second floor to see if it would be safe to put a full crew up there to overhaul. Following the stairway back to the first floor, we entered the main room beneath the bedroom. We shined our lights on the underside of the floor above, evaluating the extent of the fire damage. It was significant, and with the amount of damage to the carrying beams, it became evident that a full company could not be safely accommodated on the floor above. Then we started to look around and evaluate how much fire remained on the first floor and how extensive the overhaul operations would be. Still in the back of our minds was the possibility that the owner's body was still somewhere in the building. We knew it was impossible that anyone could have survived that fire.

As we worked our way toward the door to the rear hallway, our hand lights passed over and quickly returned to a form partially lying against the wall next to the door. Unfortunately, we now knew the location of the owner. Apparently he had come downstairs and found the fire. As conditions deteriorated rapidly, he collapsed before making it out the back door.

"Grab me a tarp, would you?" I called out the window to a nearby firefighter, as Troy advised command by radio that we had found the occupant. Troy and I then gently covered the body. There was no sense moving him yet; his position and location might help the investigation. Troy had known Mr. Black since he was a kid, so it was tough for him. One of the hard parts volunteers face is that most of the victims are our friends and neighbors. Troy kicked himself, second guessing his decisions on the attack. Could he have done something better, faster, or different? I told him the story of the house off Route 107 where I'd been five feet from the victim but didn't know it, and how I'd second guessed myself as well.

"Look at the volume of fire you had on arrival. He was probably gone before the first phone call was made," I told Troy. He knew that, but it didn't matter at that moment.

There was much more work to do. As much as we might have liked to, we didn't have time to stand around and talk. With the body well-protected with a tarp, overhaul continued and after about an hour of ass-busting labor, the remnants of the fire were out.

After the coroner had come along with the Pennsylvania State Police Fire Marshal, it was time to take the body out. He was carefully carried in a body bag to the front yard and placed on a stretcher from the coroner's hearse.

I was glad that Mike had slept through the call. He didn't need to see this end of the business yet.

• • •

Rural or suburban, fire makes no distinction, but many times when the country departments get called to the more developed valley, it's because water is an issue. They can't get enough water out of the hydrants and we can bring them lots more. Typically we would get called and then set up and run the tanker shuttle operation for them; that was the task on this early summer morning. Starting down the hill out of the township, the glow in the sky to the east was an indicator that we'd be using lots of water. It looked like an early dawn. I called command a couple of times to find out what they were going to want, but couldn't get through. There was so much radio traffic with units talking over one another that I just put the mic down. No sense adding to the chaos.

I parked on the bridge well away from the scene and marveled at the flames two hundred feet in the air from the massive piles of burning pallets. I tried to reach command again on the radio, but still couldn't get them. I decided it would be easier to try to find somebody who could tell me what they would need in person. Putting on my gear, I could hear our engine coming. Still monitoring the radio, I heard them given at least three unique sets of orders, all different and conflicting. As the engine came over the bridge, I stepped out into the roadway to flag them down. The pumper pulled to a stop, and I moved to the officer's side. The captain, Jerry, was riding the seat. As I walked up, Jerry hollered down what his most recent set of orders was.

"Don't worry about it," I told him. "We're not going to contribute to making this any more of a cluster fuck than it already is. We're going down in there and you're gonna do nothing until I can find somebody who knows what the fuck is going on."

I climbed up into the right jump seat and stood, holding onto the hand rail. The driver started down the entry road to the pallet plant and looked for an out of the way place to park temporarily. Once he found one, I jumped down and went looking for whoever was running the thing. Searching for a white helmet, I tried to slip between the burning pallets, a good seventy-five feet away, right alongside a two story building. It was so hot I had to cover my face in order to make it. As I trotted through the gap, staying next to the building, the second floor lit off over my head, just from the radiant heat.

I didn't find the chief. I found Rudy instead. He was the assistant chief and a good guy, but he was even more frustrated than I was at that point. He wasn't in charge, formally, but he was trying to get things organized.

"Don't light a match near the chief," he told me. "He's blistered." I rolled my eyes and tried to figure out how to process this wonderful information.

"Are we working for you right now?" I asked.

"Yeah," he replied.

"Okay, what are you going to want us to feed?"

"How about Ladder 20 and that gun," Rudy motioned to two pieces setting up.

"Do I have water supply and can I call direct for tankers or do you want to do it?" I asked.

"It's yours and you can call direct," he authorized.

We discussed where we were going to fill from and who was available to do it. Rudy then went off to deal with the other issues that were now his responsibility thanks to his inebriated boss.

I did some rough calculations in my head, and then keyed the mic on my portable.

"Lackawanna, Chief 36B," I called.

"Lackawanna by," came the response.

"Chief 36B has been designated water supply for the tanker shuttle and been given dispatch authority," I started. There were two tankers coming, ours and Greenfield's, but that was nowhere near enough for what we were being asked to do.

"Give me tankers from 28, 63, 18, 2, and Waymart to start," I told the dispatcher. He acknowledged and the tones started dropping. That gave me seven to work with initially. I had the captain start to set up the dump site using our engine and told him that would be his. One of the Waymart officers was given the fill site, one exit up off the nearby Casey Highway. It took awhile to get things set up as some of the tankers had at least a twenty minute run to get to the scene.

Things started coming together, but it was obvious we were going to need more tankers for the flow required and the distance involved. Getting a hold of somebody's cell phone, I called the County 911 center. After identifying myself, I asked the dispatcher, "Who would the four closest tankers out of the Poconos be?"

"I'm not sure. Hang on," he responded and the phone clicked as I was put on hold. Ten seconds later he was back with an answer.

"Okay, give them to me," I told him. Click again, and I hung up as well. I could hear their tones start sounding over the dispatch frequency on my portable.

In the meantime, the sloshed chief was driving around the fire ground in his minivan, issuing contradictory orders hither and yon. One of the county EMA officers, known as Hot Sauce, was on scene with the Field Comm Unit, and he called a brief meeting of the various command officers running important aspects of the operation.

"Look guys, we've gotta do something about him before he kills someone," he began. We discussed a command structure which would take him out of the loop, giving Paul, the Mayfield deputy chief, command. Rudy would handle operations. I would keep doing water supply. Once decided, Hot Sauce called County on his cell phone.

"Here's the deal and don't ask any questions," Hot Sauce instructed the dispatcher. "Accept no orders from the chief that haven't been cleared by the Field Comm. Just acknowledge him and then call us. Understand?" Just like that we had taken the fire away from him. It was highly irregular, but necessary in this case.

Once that unpleasant task was done, things became more controlled and organized. Then it was simply a matter of pouring water on it. A few hours later, enough fire had been knocked down that we could discontinue the tanker shuttle. The public water supply alone would suffice to finish things. With the eleven tankers we used, we established and maintained a fire flow over one thousand gallons per minute for hours, not an insignificant task.

Rumor had it that our buddy—the chief—later denied even being at the fire. Amazing what a high blood alcohol level can do to a person's memory.

• CHAPTER SEVENTEEN •

Better to Be Lucky Than Good

T he address was close—only a couple of houses up the street. The dispatcher said a man was pinned by a tree. On the way, I saw a couple of guys working along the road and stopped. They said they heard screams coming from the adjacent house, but they had-n't thought much of it. I was stunned that they made no attempt to find out why someone was screaming. This isn't New York City.

I turned around and pulled into the driveway as a girl ran toward the car waving frantically at me, a phone to her ear. I recognized Laura, my neighbor Ed's daughter, still talking with the county dispatcher. Grabbing my kit out of the car, I followed her to the rear of the house. The tree was down on the patio, limbs and branches creating an obstacle course between us and Ed.

He was lying face up on the stone, but as I worked my way to him, I could see he wasn't pinned by the tree. Reaching him, I realized it was the only good news. The stone beneath him was wet with blood, and the puddle was massive, covering at least four square feet.

Laura was still talking to the dispatcher saying, "Oh thank God, Mr. Ryman is here." She looked at me, terror across her face. "Do you know how to do CPR?"

Breaking branches away so I could gain better access to him, I could see he was still breathing. I asked her for the phone and identified myself to the dispatcher.

"Tell the unit and the ambulance both to expedite," I told him. The word expedite in our parlance means the situation is as bad as it gets.

"Okay," he responded. "Do you need the rescue?"

"He's not pinned, but keep it coming for manpower," I replied.

The laceration on the back of his head was big. I couldn't see it since we couldn't turn his head and neck, but I didn't need to see it to know the extent. There was brain matter in the blood alongside his head. I pulled the largest dressing I had out of my kit, and worked it under his head. Laura held it there for me. While she held the pad I checked over the rest of his body for injuries. Nothing significant jumped out. He had snoring respirations which are just what they sound like and are not a good sign with a massive head injury. I really wanted a suction unit to clear his airway, but it wasn't something I carried around in my back pocket.

The township cops arrived. They stood by helplessly as well. By this point, I had done everything I could with the equipment available.

"Tell the ambulance I need a suction unit, long board, and straps, quick as they can," I told one of the cops. He got on his radio and relayed the request.

The wait was excruciating. Five minutes felt like an hour.

The ambulance finally arrived with the equipment. I fired up the suction unit and cleared out his mouth and throat. Rick, another firefighter, slipped a plastic oral airway in and we started supporting his respirations with the bag mask.

We cleared more branches away to lay the long board down and carefully slipped Ed on, strapping him quickly to the heavy plastic. Surrounding him, we lifted the board together and carried him to the stretcher, situated around the corner of the house, and strapped him to that. Together we carried the stretcher to the back of the rig where the paramedics were setting up their equipment. The back doors slammed and they were gone. I figured Ed would be, too.

Arriving at the trauma center, the crew wheeled him in. The doctor took one look and shook his head. It was over.

The tree couldn't wait, Ed's daughter told us. He had insisted it had to come down, and wouldn't wait for help. When he sawed through, it kicked, and he ran, but not fast enough. It got him in the back of the head.

Back in my office, now meaningless paperwork still sat on my desk. I tried to dig back into it, but couldn't. I was just too sad for Ed's wife and daughter. Nobody likes to watch their neighbor die. Every time it happens it just reemphasizes the point that our victims are our friends. It's something many urban firefighters don't have to deal with.

• • •

Sometimes the results are much better. One day, we were dispatched for a wreck at the beginning of a road construction area on the interstate, right near an exit ramp. As I came across Route 524, I could see the highway bridge now looked like a parking lot. Driving up the off ramp against traffic, red light on the dashboard flashing, I found a spot near the top to park.

The troopers who had been monitoring the work zone were on scene. One of them was rather agitated. Actually, he was royally pissed off. Two other troopers were trying to keep him away from one of the nearby tractor trailer drivers. It didn't take long to understand why. The first tractor trailer was stopped waiting to merge from multiple lanes to a single lane in the construction zone. The second tractor trailer didn't stop and the cab was smashed into the rear of the first trailer. The problem, and the reason for the trooper's bad mood, was in between the tractor trailers. And it was about to become my problem.

A car with two girls had also been stopped behind the first tractor trailer. The second one hit the car and drove it—and the occupants—beneath the semi in front. Our pissed off trooper had witnessed the entire thing and was totally distraught.

With two rescues coming, we had plenty of tools, but what we desperately needed were ideas on how best to use our tools in order to get the girls out. We got the second tractor trailer moved back a few feet. It was damaged, but not to the point where it couldn't move. We stabilized the girls as best we could with collars through what was left of the side windows beneath the crushed roof while hooking a wrecker up to the car. We had to get it out from under the first trailer to have enough access to extricate the young women. Slowly the wrecker cable tightened, and with metal screeching, the car was pulled inch by inch from the space beneath the trailer.

Once out in the open, the firefighters made quick work removing the crushed remnants of the car roof with Hurst tool cutters, taking each roof post in one bite. The passenger side door was next to go. That gave us good access to one girl who we quickly placed on a back board and carried to one of the waiting ambulances for transport. That was the easy part. The second girl was pinned much worse, only her head and shoulders were exposed. We went to work on her.

"What's your name?" I asked her.

"Julia. Is my friend okay?"

"She's going to be fine," I tried to assure her. I had no idea how her friend was doing.

It was hard to discern the extent of her injuries due to how badly her body was pinned. We requested a helicopter to speed her trip to the hospital once we got her out.

Starting with the dashboard, which pinned her legs and abdomen, a hydraulic ram was inserted on the driver's side and a cut was made near the base of the front or A post to allow the dash to roll. As the ram began to extend, a fitting blew, and hydraulic fluid began to spray out. The power unit was immediately shut down, stopping the spray. "Goddamn it," I said. The crew looked at it with an eye toward repairing the connection on the ram. We didn't have time for that crap.

"Get that fucking thing out of here, and get the other one," I ordered.

The second ram worked, and the dash moved six inches or so, gaining us some ground. We worked with a Sawzall, and other hand tools to remove the plastic components sandwiching her in the car and, bit by bit, we made progress.

The victim was a sweet girl; she kept asking about her friend.

"She's fine, Julia," I answered her more than once. It also made us wonder if she had a head injury.

One of the troopers leaned over from near the right wheel well and held her hand. He kept reassuring her as well.

"These guys are the best on the highway. They'll have you out of here in no time," he lied to her. It was nice to hear, but we were running out of ideas. I knew it and so did the trooper.

We rolled the dash from the passenger side. Another couple inches were gained. More cuts with the sawzall; another inch gained. We were starting to get a better idea of how she was pinned as well as the severity of her injuries.

The noise from the tools was drowned out by the slapping rotors of the incoming chopper which landed on the highway just north of the wreck. We could get her to the hospital in mere minutes if we could just get her out of the wreckage. With the extra access we'd gained, I could see that there were some wires tangled between her legs. I could just reach them with some wire cutters. I leaned in from the hood, burying my head beneath the dashboard and snipped away. Her legs were free! The other guys cut a few more pieces of plastic and we had her.

We lined up firefighters on both sides of a back board and, with a medic holding traction on her neck; we carefully slid her out of the wreckage onto the board. The forty-five minutes it took to get her out seemed like eight hours.

Exhaustion set in; not from the physical labor, but from the stress. I was whipped, my clothes under my bunker gear were soaked with sweat. A bottle of water went down in one long swallow.

I watched as the bird took off with our patient, who I learned later made a complete recovery. The trooper who had been holding her hand throughout the incident came over and threw his arm around my shoulders.

"Chief, I have no fucking idea how you got her out of there."

I laughed and looked at him. "You know what? Neither do I."

He laughed as well, slapped me on the back, and we went our separate ways.

• • •

The old cliché—*it's better to be lucky than good*—sometimes combines with—*being in the right place at the right time*. It was mid-afternoon, and the old farmhouse was well involved. Clifford, a department north of us in Susquehanna County, had made a good attack with the manpower they had, blitzing with the deck gun from the driveway and knocking down the bulk of the heavy body of fire. Still, a lot of fire on both floors remained. The incident commander developed a plan to simultaneously put crews on the first and second floors to complete the extinguishment and then begin the overhaul. We talked about this a bit, and decided it would be more prudent to deal with the first floor initially, so we could get a better look at the structure and make sure it was safe to put guys on the second floor. The stability of the second floor was already questionable in my mind, as the stairs going up were gone, burnt away.

I took a team of four to the first floor; two guys on the line and two with hooks and tools opening up the walls and ceiling. They were making good progress on the remaining fire and I was starting to get a look at the supporting elements of the structure when we had a bit of a surprise.

I was kneeling in the living room when a massive crash occurred. Visibility instantly turned to shit. Initially, all I could see was that a portion of the second floor had let go. I ordered an immediate evacuation and started counting heads. "Everybody out!" I yelled through my face piece. The radio mic on my shoulder started screaming with officers outside calling for status reports. I ignored that, at present, pushing guys out the door until I was sure they were all out. About then, the smoke started to lift a bit, and I looked over about two feet to my right. There was a freezer sitting there, which thirty seconds before, had been in a room on the second floor. I exited the building and told the incident commander and the other officers outside what had happened. My hands shook a bit, and my heart still pounded. I was more relieved than scared; relieved that everyone got out alive.

If I had knelt two feet over to the right, I would have been just another line of duty death statistic. Maybe it's not a cliché after all.

• • •

Revenge has its own clichés and many arson fires seem to be the result of boyfriend/girlfriend disputes. This one was no exception. We didn't know any of that in the beginning; on the way to the call, things seemed more than a little confused. First, there were conflicting stories as to whether or not someone was still in the building. Apparently, the first chief on scene was arguing with a resident who refused to evacuate. Second, it was reported that the fire was knocked down, followed by a report that it wasn't.

I decided to stop paying attention to the radio transmissions and size everything up from scratch upon arrival. I parked, put my gear on, and grabbed an air pack off a nearby engine. The house was a split-level ranch. There was heavy smoke coming out the front door, so the report that the fire was knocked was obviously incorrect. Peter, the chief, and another guy were starting in the door with an inch and three quarter line.

I walked up to the door, knelt down, kicked my helmet back and put my face piece on. With air now flowing, I half nudged and half shoved my way past some guy from our sister company who was kneeling in the doorway doing nothing but sucking air. I followed the line to the top of the stairs where I could feed it to them as they advanced down the hallway toward the bedrooms. The line stopped, and I felt the surge as they opened the nozzle. I could hear the stream impacting the dry wall in the distant bedroom, like a drummer pounding the tom toms, as the water went from wall to ceiling and back again.

A minute later, I heard their bells on their air packs go off, and they started back down the hall toward me. They must've been on air previously doing something else, because we had only been on this attack for a couple of minutes, not long enough to go through an entire tank of air. No time to worry about that now, though.

When they got back to the top of the stairs, Peter stopped for a moment.

"It's not out. I could hear it still crackling in there," he said.

"Okay, get me another guy and a camera," I told him.

They had left the nozzle all the way down the end of the hall. I was by myself, and had no idea, in the heavy smoke, if fire was rolling down the hallway ceiling. I needed that camera to determine if the crawl down the hall to get the nozzle was possible. I didn't want fire behind me. I didn't want to try pulling the line back to me as the pistol grip on the nozzle had a tendency to catch on anything in the way.

In a matter of moments, a guy came in with the requested thermal imaging camera. Checking the hallway, we found it was hot, but free of fire at present. Down the hall we went, crawling on our hands and knees, to get the nozzle. Reaching it, I turned the nozzle into the doorway of the bedroom, still on my knees. The opposite side of the room was burning with fire rolling across the ceiling. We didn't need the camera to see this.

I opened the nozzle on straight stream, directing it at the ceiling, whipping it around, creating large water droplets to work on the fire. The windows were still intact; without ventilation, the knockdown was taking longer than it otherwise would. I tried, unsuccessfully, to find the windows with the heavy stream. If I had hit one, it would've shattered and helped the smoke and heat to lift. I wasn't sure why we weren't getting ventilation from the exterior yet, but I couldn't do much about it at that point.

Ultimately, we were flowing more water than the fire could withstand, and it went out. I told the kid behind me to have someone outside take out the bedroom window. We still couldn't see it in the heavy smoke. He crawled back to the stairs and relayed the message. A few minutes later, there was a crash, as the glass broke and the smoke and residual heat quickly lifted. I gave the room another quick wash down with the hose line. The fire was knocked, a few hot spots remaining for overhaul.

The torch was clever. He had started the fire in the basement, below the bedroom, near a bunch of wires. It spread up through the floor into the bedroom. It was designed to make it look electrical in origin. It took the Pennsylvania State Police Fire Marshal a couple of visits and some interviews before she figured it out. Initially, it fooled her, not to mention everyone else.

The guy the chief had been fighting with to get out of the building was trying to put the fire out in the basement. The house belonged to his girlfriend, with whom he had been fighting. He was our torch, and apparently had changed his mind. That change of heart, however, didn't save our reluctant arsonist from a stay in the pokey.

• CHAPTER EIGHTEEN •

A Flick of the Switch

My son Mike was thirteen-years-old when his interest in firefighting developed becoming the third generation in a family of firefighters. To me, his transformation seemed instantaneous like the light coming on at the flick of the switch. He began tagging along with me on calls that I chose selectively for time of day and type of call. I didn't want to expose him to anything really ugly—inevitably that would come if his interest remained.

He started by learning how to rack or reload hose as well as how to change an air cylinder. He picked up the names of various pieces of equipment, both real and slang. I didn't push him. I've seen too many sons join because of their fathers. Either they were forced to sign up or they joined out of some sense of obligation. Most of them, the sons that is, were worthless as firefighters—they really didn't want to be there and it showed. You can't manufacture the desire to do this job.

Mike couldn't wait for his fourteenth birthday which was the required age to submit the paperwork to join as a cadet. He still wouldn't be allowed to do a whole lot but he could increase his knowledge by taking a few classes. The timing was good as I had hung my white helmet up for good. Now I could concentrate on working with him and observing him in the field. After he joined, just a few weeks went by before he got his first official lesson. He'd only been on a few calls, and none too serious. It was still all cool-looking gear, flashing lights and blaring sirens to him, regardless of the wisdom I tried to impart.

A wreck came in on Heart Lake Road in the northwestern portion of Scott Township, a vehicle on its side with entrapment. We went to the scene and found an SUV on its driver's side with a woman in the front. She wasn't hurt badly, and was actually standing up; her feet were on the driver's door.

One of Mike's friends, Dan, was standing next to the car looking through the windshield. The woman inside was his mother and the wreck was actually in front of their house. We stabilized the vehicle and I popped the rear window to get the EMT into the vehicle and he worked his way up to the woman to evaluate her. Mike helped haul equipment off the rescue, but out of the corner of his eye, kept watching Dan, who was obviously upset about his mother.

The EMT got a blanket over the victim and we cut out the windshield to gain access. A collar in place, we laid her onto a back board and secured her with straps, all very routine. Mike helped place her on the stretcher and then carry her to the ambulance. Dan got in the front seat of the rig to go to the hospital with her. Mike could tell he was worried as they left for the emergency room.

We picked up the equipment to get the rescue back in service. On the way home, we talked about the call.

"When you do this job in a rural area like this, there are many times when you'll know the person involved. That's not fun or easy, but it's part of it," I told him. It was the first call he'd been on where he knew someone, and the first time he realized these things can affect friends.

Dan's mother wasn't badly hurt, but it was a good learning experience for Mike.

• • •

The learning wasn't over. In fact, it never really stops. On his next call, the initial reports were for a structure fire. They were both right and wrong. It started as a brush fire that extended to a small out-building, and a bit to a barn. The first arriving engine quickly knocked down the small amount of extension to the barn. The small shed was pretty much gone. The brush fire portion wasn't going anywhere fast, but it had gotten into some of the farm storage and the trash that was in the yard, which meant the overhaul would be a pain in the ass.

Mike was all excited. I was happy for him and moaning inside for myself. If it weren't for him, I sure as hell would not have been happy to be there. We worked our way around the back and took the line that was working the remnants of the out-building, giving the original firefighter a break.

I gave Mike the nozzle and told him to brace himself and open it slowly. Despite my warning, his unskilled hands opened it a bit too fast, and it almost got away from him. Quickly, he slammed the bail shut. The second time, he was more prepared and was able to start working the remaining fire. I pushed forward on the line behind him, absorbing some of the back pressure as I directed him. I could look around his helmet and see his eyes glowing and the massive smile on his face. It made being on this shit call—ass busting work, but an easy, routine fire—completely worth it. I was watching my son on a line for the very first time, and I got to be right behind him. For a minute or two, I thought I was going to cry.

• • •

The easy stuff didn't last long. On our way to a call for an accident with possible entrapment, the few reports coming over the radio from the dispatcher were beginning to hint that this might be a nasty one. Mike hadn't been on a bad call yet, but when we turned the corner onto Route 524 near the Scott exit of Interstate 81, I was pretty sure that was about to change.

As we pulled into the gas station, I could see the cab of a tractor trailer buried in the side of the building. Exiting the car, we put on our gear. I started to size up the situation and evaluate my initial concerns. The first thing that became obvious was that power was still on in the building and in the truck. With the heavy damage to both, this could be an ignition source. The extensive structural damage to the building was the second issue; so heavy that the integrity of the building looked questionable.

Mike walked around the corner of the truck ahead of me. When I caught up to him, he was standing, staring at the truck. I saw what had his attention. A woman was on the ground partly under one of the trailers wheels. It was pretty obvious she had gone through the wheel well. The word meat grinder came to mind.

The woman's husband was standing there staring blankly at the carnage, obviously in shock, with some lacerations to his head and face. I tried to get him to sit down in a nearby police car, as other firefighters began to arrive. The truck driver was reported to be out of the tractor, but no one knew where he was.

Jerry, an assistant chief, and I entered the building to start a search. We received a radio call that the driver had been located outside, but we still didn't know if anyone else was inside. The first thing I did was look at the secondary collapse possibilities. The news was good...sort of. I could see that the tractor was doing a really good job holding up the building. The guys outside worked on getting the power killed. We started to climb an eight foot high pile of debris, a mountain of crushed shelves and junk food, to reach the center of the building where the tractor was. When we reached the top, we heard someone cry out from a nearby pile of shelves.

Climbing down the other side, we found a teenage boy pinned by three shelving units. They were the wire and steel models common to most gas stations. My initial thought was that he was impaled, but after clearing away some of the debris, I could see that he wasn't. I picked up Jerry's radio and told command we had located a victim and were going to need cutting tools—bolt cutters, hack saw, Hurst tool cutters, Sawzall—anything that would cut.

In addition, I told him to have a crew start clearing a path to us from the B side of the building. There was no way we would get this kid out the way we had come in. Extra manpower inside wouldn't help; we were surrounded by piles of merchandise and other shelves. The boy's father was in there, too, but uninjured. There was only room for Jerry and me to work on the victim.

"Help him, please," the father screamed at us. He was yanking helplessly at the three dimensional metal jigsaw puzzle encasing his son. While we understood his reaction, he wasn't helping matters.

"You've gotta let us work. We won't leave here without him," I said pulling him out of the way. We started clearing away more of the merchandise from the shelves pinning the young man while awaiting our tools.

The first thing to reach us was a hack saw. I started cutting some of the shelving cross pieces while Jerry kept clearing away the debris around us. I could see it was going to be a long day if other equipment didn't get there soon. At about that time, a pair of bolt cutters arrived and we began to make quick work of the wire shelves pinning the boy's torso. The boy's piercing screams only increased the tension for us. We wanted to move fast, but we had to proceed with caution so as not to injure him any further.

Meanwhile, a crew began a bucket brigade, clearing shelving units, merchandise, building debris, etc. to make a path for us. Mike brought the heavy Hurst power unit to the side of the building and set it up, connecting the hydraulic hoses to the tool before starting the engine. Usually this unit is carried by two guys because of its weight, but there was enough adrenaline pumping through his young body that he took it by himself. The cutters were passed in to us and I began using them on some of the heavier shelving components while Jerry kept working with the bolt cutters on the lighter parts. For some reason, the sawzall was no help. It was like using a butter knife on granite.

Out of the corner of my eye, I could see Mike was working the bucket brigade of debris. By that time, we had freed the boy down to his knees. The sweat was pouring off me and my arms were getting sore. We kept clearing and cutting until we reached his ankles. There was only one stubborn piece remaining. His foot was wedged beneath the lower heavy metal shelving bracket.

With two final snips of the Hurst tool cutters, he was free.

A Reeves stretcher was waiting for him. Jerry had the boy's arms around his neck, and he passed him to me. I took him and laid him downward onto the Reeves. His ankle was broken, his foot twisted to the left at a ninety degree angle to his leg. As gently as we could, we wrapped the Reeves around him and passed him to the waiting crew, which included Mike, who carried him out to the waiting medics.

Jerry and I staggered out and half collapsed onto a blue tarp which had been a staging area for equipment. My forearms felt like I had spent the last eight hours doing curls at the gym. I sucked down a quick bottle of water and got ready to go through rehab where we would be checked over by the EMS crew. First, I wanted to check on Mike. I was glad to see he was fired up and happy that we'd been able to make a good rescue. He was able to put the ugliness of the fatality in perspective. Overall, it was a positive experience and a big step in his development as a firefighter.

When we got home, he headed for the shower while I went to talk to Michelle. I usually didn't tell her much about calls like this. She didn't want to know. But with Mike involved, I thought it might be a different matter.

"He's gonna need to talk about this," I told her. "You may have to hear about it."

She understood and steeled herself for when he got out of the shower. Then I got one of my better ideas—my father would love to hear about this. Mike got out of the shower, all ready to burst.

"Hey, why don't you call Grandpa and tell him about the call?" I asked. Mike thought it was a good idea and talked to my dad on the phone for a good half hour. It got his mother out of hearing the ugly stuff. This time.

• • •

He certainly told his Mom about the next call getting me in hot water in the process—a feat Mike enjoyed immensely.

I was sitting at the archery shop having a cup of coffee. Megan was babysitting the owner's kids. I was dropping her off, but decided to stay and chat for a few minutes. Just as I sat down with my coffee, the pager went off for a house fire about two miles down the road.

Jumping in the car, I headed in that direction. I pulled over edging off the road in order to leave enough room for where the apparatus would set up. As I got out of my Explorer, I heard a bloodcurdling scream. I didn't take the time to put on my gear as I grabbed the fire extinguisher from the back. I trotted over to the side of the house and saw the owner along with his wife, who was hysterical. I was expecting something much worse.

"Anybody in there?" I asked.

"No, everybody's out," the man said.

"Where's the fire?"

"It's in the first room on the right."

The smoke wasn't really bad, so I crouched, and duck-walked in. The door to the room was hot, but not exceptionally so. I cracked it and no fire came blowing out at me, so I opened it a bit further. Fire was running the wall behind the door starting across the ceiling. I pulled the pin on the extinguisher and squeezed the handles. Dry chemical agent began to flow and I directed the hose at the seat of the fire and then up the wall to the ceiling. When the room went black, I quickly slammed the door shut.

Now the owner was in the hall beside me.

"Do you have a garden hose?" I asked. He quickly returned with the hose.

"Turn it on," I told him. Soon the hose came to life. I opened the door and directed the stream at the ceiling, whipping it around, trying to make drops, and then behind the door where I knew the seat of the fire had been.

It looked good, but the smoke was getting thicker in the hall-way. At about that time the chief arrived and came in behind me.

"I've got it," I told him, "but we need a crew and a line on the floor above to check for extension. I'm not sure if it got in the wall." He made a quick radio call giving the orders.

"How are you?" he asked.

"I'm fine," I responded.

"Real good job—and I ought to kick your ass," he told me. I knew what he meant. I'd have said the same thing in his shoes. It was a good stop of the fire, but I was in there with no gear, no breathing apparatus, and a damn garden hose in my hand.

Michelle had brought Mike down after hearing the call and telephoning the shop to learn that I had responded. He worked on racking hose and getting equipment back on the apparatus. After I took a little oxygen, I brought him in to see the fire dam-age and to talk a little bit about the fire spread and its behavior. It was fun teaching him some of the basics. I think he was im-pressed with the barehanded stop the old man had made. His mother, however, was not.

• • •

When he was about fifteen-years-old, Mike showed he did-n't mind a little heat. It was a snowy weekday morning. School had been cancelled with the six plus inches that had fallen overnight. A call came in for a trailer fire in a congested court about a mile from the station.

Mike and I rode the jump seats on the rescue and I put on an air pack. The column of smoke was visible as we made our way through the narrow roadways inside the trailer park. The homes here were very close together. Pulling in we saw a trailer with flames already blowing from the windows.

I grabbed the nozzle of a pre-connect and Mike helped me stretch the line toward the far doorway of the burning trailer. I was the only one with a pack on, at least the only one who would go inside. Some guys will put one on, but it's mainly decoration as they have no intention of really using it. There were a couple of those with us that morning.

As we got to the doorway, the line went hard with water. I knew I couldn't go far since Mike, as a junior, couldn't go inside. I wanted to at least get just inside the doorway and hopefully cut the fire off there. Opening the door, I knew it was not to be as fire boiled out at us. It was a bit warm where we were between two closely spaced trailers. We stayed low, but the heat was intense, melting the siding on the unit behind us as we listened to the windows cracking.

I slapped the bail back on the nozzle and tried to drive the fire back into the doorway but, initially, the heavy flames just ate up the straight stream like a garden hose in the mouth of a dragon. After a half minute, the fire began to recede and I could move forward. I glanced back to see how Mike was doing. He was right with me, humping the line and taking up the back pressure, a big smile on his young face. He didn't seem to mind the heat in the least.

I made it to the doorway but went no further. The combination of Mike being my back-up and the obvious loss of the building deciding things. With additional help making it through the snow, we eventually knocked the fire down and finished it off, the building a write-off. None of that mattered to me. Seeing my kid, with such limited experience, stay right with me, had made the morning a complete success as far as I was concerned.

• • •

A few weeks later, we were dispatched to another trailer fire, this time a few miles west of the interstate on Route 374. Mike and I started up the highway and then down the twisting country road, looking for the typical column of smoke in the air. We never did spot it, but arrived on scene shortly after the first engine. Grayish smoke poured from the trailer, hanging low near the ground, but there was no visible fire. It looked like we might be able to make a stop on this.

As we finished putting on our gear, Clifford's rescue came over the rise pulled past us and parked. I walked over to it and pulled an air pack out of the rear and started donning it. I saw Ross from Fleetville out of the corner of my eye and told Mike to go hook up with him and get some work.

Walking up to the engine in the driveway, I saw Troy, the incident commander, with a head set on standing at the top mounted pump panel. I was really looking forward to doing some routine firefighting, but it was not to be. He motioned me over and told me he wasn't getting any good information and requested I take control of the attack. He had one of his rookies get me a radio, and I walked up to get an idea of what was happening. Two lines were at the trailer, one in each door. The line team from the front had already made a good knock on the fire and the other crew was just going in the other door to work toward them.

As they started to turn, they had to move a microwave and shelf out of their way. Underneath was the trailer's owner—an elderly lady. I stuck my head in to look and saw she was definitely gone. A runner was sent over to Troy to tell him and request he contact the Pennsylvania State Police Fire Marshal and the coroner, so we didn't have to put it over the radio.

I got things organized and made sure the crews were finishing off the fire and completing a search of the entire trailer. Mike was doing some exterior ventilation and worked on opening the trailer skin for overhaul in the area of the living room where the bulk of the fire had been. He had a great time ripping and tear-

ing the burnt structure apart. I called him over to the rear doorway to show him what had happened to the owner. It was his
first fatal fire.

The bed in the next room looked recently used. We suspected she had been taking a nap when she was awakened by the
fire. Unfortunately, the smoke was already bad enough that she
only made it to the hallway, and apparently grabbed the shelf
with the microwave on it as she collapsed. She was three feet
from the door. I looked through the entire trailer—not one
smoke detector. The old lady died for want of a single ten dol-
lar smoke detector, something we still see too often among the
poorest of our neighbors.

• CHAPTER NINETEEN •

Billy the Kid and Uncle Mike

Early on a snowy Sunday morning, while we were snug in our beds, the pager went off and a full structure assignment for a furnace explosion was announced. The roads still had not been plowed, and snow was coming down heavily when Mike, just days short of his sixteenth birthday, and I headed out.

Naturally, the house was at the top of a long, steep drive-way—they're only close to the road when it's seventy degrees and sunny. I put the vehicle into four wheel drive low and up we went. We didn't even spin a tire.

There was nothing showing smoke or fire-wise, but the house had sustained tremendous damage. The chief arrived on scene and we started to size things up as best we could in the snow-filled darkness. The two walls of the first floor master bedroom had blown out, which was probably what saved the occupants who were sleeping in this room when the furnace exploded, as it relieved the pressure. The rest of the house had been moved from the foundation. Windows—frames and all—had been blown from their wall openings. We were stunned by the level of damage, and yet there was no fire.

The family dog had been sleeping on the floor next to the bed. He was now dangling from the second floor rafters, his head wedged between the beams. Mike's eyes went wide at the sight. It had been quick for the dog; I don't think he knew what hit him.

Apparatus now on scene, Little Billy, a younger firefighter, and I went into the basement. Mike stood by outside, ready to get us tools or equipment. Billy took a multi-gas meter with him; we still weren't sure what the cause was. We got a few very low blips of carbon monoxide, which quickly disappeared. It could've been the exhaust from the apparatus blowing into the house. Nothing else registered on the unit. We made our way over to where the furnace had been. It was an old coal burner with a water jacket. The jacket exterior of quarter inch plate was peeled off completely, the edges as jagged as a newly opened tin can. There was no soot or evidence of a flame front. A steam explosion, we concluded. Without an in-depth examination, we couldn't find an exact cause.

We helped the family salvage some personal effects and valuable items, taped off the now structurally unsound building, and went home. Luckily we were able to get the township road crew to plow and cinder the driveway, making the trip back down a bit safer. Still, it was not a great way to start a Sunday morning.

• • •

Great people, some of which become like family, always make for good days. Mike grew up seeing Guido or Uncle Mike, as he was known, on a fairly regular basis. He was the big olive-skinned guy with the huge smile and an infectious laugh; a teddy bear of a guy who cooked barbecue chicken and set off huge quantities of fireworks on the Fourth of July. His early forties comb-over gave way to a fully shaved head within a few years.

One time when Mike was a baby, Uncle Mike was delivering pizza as part of a fund raiser. He delivered ours and proceeded to play with Mike while we ate. By the time he got back to make his next delivery, they were done for the evening. The fund raiser had a few more nights to run, but he wasn't allowed to deliver to our house after that. Uncle Mike still has a hard time with Mike being old enough to fight fires. His head shakes every time he sees the kid in bunker gear. That was the Guido that Mike knew for the first fourteen years of his life. He got to see the other one late one night on a fire in Fleetville.

A chimney fire had extended to the wall and the construction was a bit unusual. The decorative false walls around the mantle created concealed spaces for the fire to run through. We were having a hell of a time chasing the fire down, and figuring out all the possible avenues of travel. Mike was sent in to give us one tool or another and then stood back to watch. Guido and I were working on opening up while having the standard, reserved, business-like discussion that takes place in such situations.

"What the fuck? You think we should open this fucker up?" Guido asked. "Goddamn this fucking thing is running. We gotta get ahead of the son of a bitch."

On and on went the typical back-and-forth that goes on when we get frustrated digging out a fire like this. After a half hour we were satisfied we had it all, and we picked up to return to service.

On the way home, Mike was quiet. He asked a couple of questions, trying to learn about the tactics and methods we had employed. Finally the comment I had been waiting for came: "That wasn't the Uncle Mike I knew."

• • •

Uncle Mike didn't just put out fires. One nice spring Saturday afternoon, Michelle and I were on our way to the movies. It was the last showing of a particular film, part of a festival at a small local theater. We were cruising along Route 107, chatting about the movie, when I saw a column of smoke in the distance. It looked like it was in the vicinity of Guido's house.

"This is not good," I told Michelle.

"Mike's probably just burning some trash," Michelle said, ever the optimist. Looking at the smoke, I knew better.

"I don't think so. This is not good," I repeated.

As we got to his driveway, I could see the fire running in the woods. And as we passed it, Mike was visible with a broom, going after it. I pulled up a little farther and made a U-turn and edged over to the guardrail. Lowering the window on Michelle's side, I leaned over and yelled out of the car. "Mike, what the hell are you doing?" I got the teeth, the big Guido grin, in reply. Obviously he had been burning trash and the fire got away from him.

"I called it in," he hollered. "Ross is bringing the brush truck."

I shook my head and pulled into the driveway and parked. Grabbing my bunker pants, I kicked off my shoes, pulled them on, and put on my helmet and gloves.

Mary, his wife, was yelling at him: "Michael, get up here and put your bunker pants on!" Two or three times she told him. He pretended not to hear.

She gave me a knowing look. "Gary, can you deal with him?" she asked. I nodded and laughed and headed down into the black. I caught up with Guido at the head of the fire.

"Gimme the fucking broom and go get your gear on. I don't think those gym shorts are nomex," I told him.

The teeth again, that amazing smile flashed my direction. There are only about three people alive who could talk to him like that, and luckily, I was one of them. Mike climbed up the bank and put on his gear while I went to work with the broom. I wasn't making much progress. It spread faster than I could handle with the broom. Fortunately, I heard the siren of the brush truck with Ross coming down the road. Climbing up the bank to the road, I flagged him down. I wanted him to stop there so we could knock down the head of the fire with the booster line.

He pulled to a stop and I grabbed the nozzle, hose peeling off the reel as I pulled it with me into the brush. Ross started the pump and charged the line and I started working my way down the shoulder into the head of the fire, killing it as I went. It wasn't a big fire, but with just three of us and the wind doing weird things, we needed to knock it down quickly before it started to run.

I was in no mood, or physical condition, to chase this thing all afternoon. A couple of minutes of water, and it was under control. The hard part of finishing off all the hot spots had to be done next, one of the reasons I hate brush fires. They're like real work.

Fortunately, a few more guys showed up, including a couple of younger ones. It took forty-five minutes or so, but we finished everything off.

Then I received more good news.

Michelle had decided to move the car farther up the driveway. After she did, she got out and somehow locked the keys in the car. I just shook my head. Howard ran her back to the house to get the spare set of keys, while Mike and I had a drink of water. All the apparatus left, so we switched to a nice glass of wine. Soon, Mary broke out the rest of her Italian delicacies and another bottle.

Michelle and my son, Mike, arrived back at the scene. Mike who loved brush fires, was pissed he had missed it, and just had to come over to see the results. He walked around looking at the blackened woods, wishing he had been there. Meanwhile, we dug into Mary's incredible spread and soothed our smoky, parched throats with the sweet red wine. Not the typical way to finish off a brush fire. *Movie? What movie?*

• • •

Another important person in our lives is Little Billy. He's not little any more, but he'll retain that moniker for at least a few more years. He was a grade school kid when I first saw him. My first memory of him is from a fire in Fleetville at the infamous Goat Lady's place. The large barn across from her house was burning. After school was out, his mom had brought him over to watch the fire and his dad, Big Bill, was pumping one of the pieces. I had ridden the seat of the engine on the call, and Big Bill was going to drive it back.

Billy's mother had gone back to work so he was going to ride back with us. With the amount of sooty, wet gear, empty air cylinders, and dirty SCBA, the only place for the kid to sit in the cab was on my lap, so I held onto him as we rode back to the station.

Ten or eleven years later, or a blink of the eye, we were doing an acquired structure burn for training purposes. We always liked to bring the rookies in when we lit a room off so they could watch the fire development before the line came in. It was a good opportunity to let them see how fast a fire could develop and spread, and learn a little bit about these topics. We had one new rookie who was going to pop his cherry that night, squatting in the hall with me watching the fire start to roll across the ceiling of the bedroom and out the door. It was Little Billy's first time inside, and it made me feel old.

In just a few years, he has developed into an excellent, well-trained firefighter, officer, and a degreed engineer. He's now mentoring another aggressive and smart kid and watching the development of a young man he watched grow up—my son Mike.

The similarities between the two of them are striking, and they've developed a close relationship. I like Mike having someone from a younger generation that he can go to for answers and someone who understands the state training bureaucracy from having experienced it firsthand. The fact that I trust him and the answers he's giving my kid also helps. They now scuba dive together, and Mike does odd jobs for Billy around the house he's building. It will be fun to watch how it turns out.

• • •

The learning process continued for Mike, not always in an enjoyable way. A QRS or quick response call at the state park for a self-inflicted gunshot took Mike to his first suicide, the spring following his sixteenth birthday. The victim, a forty-something male, was found lying on a picnic bench with a bullet wound to the head, there was nothing really for us to do.

The actual scene was quickly cordoned off with crime scene tape to keep people away. There were a few who appeared to be family or friends of the deceased standing outside the perimeter. Upon leaving, I asked Mike if he had noticed them at all. As I expected, he hadn't. This type of call, I explained, is a good example of how a seemingly routine call with no real work required—although ugly—has secondary and tertiary effects.

"Think about how absolutely devastated that man's friends and family are," I said. I didn't want him to become desensitized to these situations. We go, we stand around a bit, and then we leave. Loved ones have to deal with the results for much longer; there really isn't much you can say in such a situation. Sometimes a pat on the shoulder, just a human touch, can help; other times nothing will.

Firefighters are witnesses to ugly, tragic events. The longer you do it, the more of them you see. Everyone deals with these things differently. I compartmentalize these incidents. I can't think about them constantly or even regularly or I'd go crazy. I don't have nightmares or dreams about calls. I stick these events in a corner of my mind behind a door in a room that only gets visited on occasion. I have to go there every so often to maintain my humanity, but not so often as to destroy my ability to do the job.

Everyone who does this job for an extended period of time is a very different person than they would have been had they done something else. You can't see and experience the things we do and not have it change who you are. For myself, I do think it has made me more immune to people's suffering, harder and more distant. Not because I don't care, but it's my protective mechanism. At the same time, though, it has made me more sensitive. I avoid sad movies. Simple scenes like the boy's dog getting shot or similar stories that have little or no effect on "normal" people bother me a lot. The way I look at it, I expose myself to real world tragedies. I don't want to watch a movie or television program about fictional ones for entertainment.

• • •

Sometimes, a routine structure fire is just the thing to recharge old batteries.

"Get your ass in gear," Jason yelled as he ran up and pounded on the side door of Ross's house. As he ran out to see what the commotion was, Ross instantly saw the smoke pouring from the eaves of C.J.'s, the deli next door. He called the county on his portable while doing a three-sixty to try to see how far the fire had spread. Between cursing to himself at the slow response of the moron dispatchers, he requested a full first alarm assignment, and then went in to see if he could locate the main body of fire and, more importantly, do a quick search. He made the stairs and did a search of the bedrooms on the second floor as the smoke started to bank down. He knew the fire was in the room on the left at the top of the stairs. The door and wall were hot and he could hear it crackling in there.

With the pagers activated, the rest of us were on our way. The engine pulled onto the A corner (left front) of the building and the first line stretched to the door on the B side. I pulled onto the front lawn of the house across the street. It's a good thing they're not picky about parking in the country. Mike was busy at a school activity; I knew he'd be ticked when he learned he missed a working fire.

I kicked off my shoes and jammed my feet into the boots I had dropped on the ground from the back of the car, and pulled my bunker pants up. I grabbed my coat and helmet and trotted over to the engine. Out of the corner of my eye, I saw Guido donning his gear by his pick-up. I pointed my index finger toward him and my thumb back toward me and then motioned toward the building. He nodded, divining that we would team up together and go in. We loved working together and at our, shall we say, well-seasoned age, had become known as the "Geritol Crew."

I grabbed a pack from the jump seat and carried it toward the B side of the building so I could see a bit more while I put it on. One inch and three quarter was already in the side door and a second was being stretched to back up the first. Guido trotted over, and we talked briefly to Ross about where on the second floor the fire was. We learned that Steve and Rico were on the first line. I picked up the nozzle of the second line and followed the first to the stairwell.

The smoke was heavy, black, and banking down. I couldn't see the team at the top of the stairs, but I could hear them swearing. I passed the nozzle to Guido and went up the stairs to assess the conditions. The fire was in the room off to our left. Rico was flowing water with little effect as the heat was increasing. Guido handed me a Halligan and I started opening the wall. Steve ripped dry wall away from the hole with his hands so we could see the fire a bit better.

Ross yelled in that it had vented. I had Guido tell him it didn't look like it to me, and to get the fucking roof open. We found out later that there was a slight miscommunication and it definitely hadn't vented itself. Luckily, Ross trusted our judgment and didn't hesitate to get a crew on the roof.

Steve and Rico ran out of air and Guido and I took the line. I heard the saw start up, and as soon as the opening was made, we made a left turn into the room through the doorway. Now ventilated, we pushed in and knocked the fire in two minutes. I left the nozzle flowing to cool things down.

Ross yelled up the stairs that he had fresh guys. Checking my air gauge, now visible in the lifting smoke, it was obvious my bell was going to ring any second. Guido and I backed down the stairs and handed the line over to a couple of pups. Outside, I took off my face piece and gave Ross a report on the conditions.

"It's knocked. You just need to overhaul now."

I dumped my pack and yelled at Guido when I saw him getting another air bottle. "What the hell is the matter with you? Let the kids finish this thing," I told him. Ross settled the matter by telling the youngster changing his bottle to stop. He told Guido to dump his pack.

It was a good stop, with minimal damage. It easily could have turned into a five hour marathon, though. We opened the eyes of the young hot shots showing them that the "Geritol Crew" could still handle a fire.

• • •

At eighteen, I could go through three air bottles, overhaul, clean and pack up the equipment, go home and rest for forty-five minutes, and be ready to go again. At thirty, two tanks were my limit, and it took a couple hours before I felt back to normal. At forty, I could still do two bottles, but the recovery time was now extended to the next day.

Now on the edge of fifty, I try to limit myself to one tank. I can work after that, but it's limited to some light overhaul, definitely not the heavy stuff. Once home, I head for the bathroom; not to shower immediately, but for the aspirin container, trying to head off the inevitable aches and pains. They show up anyway, and stay to visit for a good thirty-six hours.

Watching Mike is like turning back the clock. He can work for hours on a fire doing ass-busting overhaul, and he's fresh and ready to go after forty-five minutes of rest. It's a bitch getting old.

At the same time, it's important to have a life outside of the fire department, since the work can become all-consuming. I've seen many people for whom it became their whole lives, and most of those ultimately burn out. I've encouraged Mike to keep and develop his outside interests for just that reason. Hunting, shooting, and scuba diving are all great activities, and I try to make sure he doesn't neglect these in favor of the fire department. My primary outdoor diversion is fishing, followed by growing grapes for my homemade wine. A year ago I went back to school, working on a master's degree in American history. Disparate interests are important to take you away from too much time with the big red trucks.

As a volunteer, you're on call twenty-four hours a day, seven days a week—at least when you're around. I'm lucky that my bosses have always been great about me taking time off for fires. Working primarily from home, I'm not wedded to a nine-to-five schedule and, as long as the work gets done and you put the hours in, they don't really care. If working on a tight deadline, I would have to pass on going to a call; after all, putting food on the table and clothes on the kids' backs has to take priority. But those situations have seldom happened. Likewise, my bosses were always happy to play Sergeant Schultz—"I see nothing"—when riding in my red light and siren-equipped company car, trying their damndest to ignore the radio and siren control head mounted in the center. They don't ask; I don't tell, and we are all happy.

• • •

Getting to calls is not always routine, but some of the problems can boil down to genetics. Back in the 1960s, a call came in one evening for a fire at the Pep gas station in New York. Dad jumped in the car, and responded to the station. As he arrived, the first engine was pulling out to respond. He and Rodney crewed the second due engine with Rodney driving and Dad riding the seat.

Out the door they went, making the right and left hand turns that would send them down Glendale Drive toward Route 17C and the Pep station. Dad's foot was heavy on the siren pedal of the old 1957 American Lafrance pumper. A few minutes later, they arrived, confused to be the only fire apparatus at the location. Then it dawned on them that they had gone to the wrong Pep station. The one the fire was at was on the opposite end of the district. In fact, they were over the line breaching the neighboring department's area.

Dad and Rodney killed the lights and snuck their way back to the correct gas station. Finally arriving on scene with the fire out, the chief asked a few pointed questions regarding their delay in arriving. They confessed, and were subject to more than a little ball busting.

Now it's 2008. Mike had received his drivers' license a few weeks before, and had driven to the archery shop to hang out. He received a QRS call to assist the ambulance on Decker Road, and since he was close, he decided to respond. He headed across Route 107 and out Worth Church Road to Decker. Slowly cruising up and down, he could make no sense of the box numbers on the houses versus the address dispatched for the call. Then it dawned on him. The Decker Road the call was on was in Abington Township, a long way from where he was, in the neighboring department's area.

He turned around, and hearing that the ambulance was already on scene by this time, went back to the shop, saving himself from any abuse, other than a few comments from me.

Okay, I can to confess to something similar, although in my case it worked out better. One evening before Mike was born, Howard, one of the ambulance officers, and I were at the Little League field watching a game. The pager tripped with a report of a camper on fire. To make it a bit easier, Howard decided to ride with me, and threw his gear into my car.

We started over to the scene, red light flashing and siren wailing.

"Do you know where you're going?" Howard asked.

"Yep," I responded. "The Snyder residence on Carey Road."

"I don't think so," he said. "It's at the Carey residence on Snyder Road."

"Are you sure?" I asked.

"No, are you sure?"

"I was until you said that."

"Well, where are we going then?" Howard asked.

"Carey Road," I answered as decisively as I could.

"And if the fire isn't there?" Howard asked.

"Then I shut off the lights and siren, and we go back to the ball game and pretend this never happened."

Howard nodded approvingly at this plan. Luckily, we picked the right road. The fire was knocked down quickly, and we returned to see the rest of the game.

• CHAPTER TWENTY •

Coming of Age for the Third Generation

Like most guys, I have certain well-broken in articles of clothing that are so comfortable I could live in them. One such item was a brown inch and a quarter leather belt that didn't look great any longer but felt like a part of me. I had no plans to replace it—even though it was starting to wear, at least from an appearance standpoint. It still did what it was supposed to do, namely, keep my pants up.

My belt was the last thing on my mind on a warm summer afternoon in 2008 when my pager went off. The call didn't sound urgent. It was at a farm up the street.

"A twenty-one-year-old male with a lacerated leg," the dispatcher reported.

There are a thousand ways to cut a leg on a farm.

As Mike and I pulled up, it appeared that there were two kids lying in the grass fooling around but the girl jumped up and began running toward us, an expression of sheer terror on her face. We grabbed gloves and the first aid kit and ran over to where the young man was lying partly under a tractor, a bloody towel over his right leg. I knelt down by his leg and Mike followed suit at his head. I lifted the towel to see how bad it was. My breath caught in my throat.

Oh fuck. His leg was gone from about four inches below his knee. There was some skin and muscle laying loosely to the side and sharpened white bone sticking out where his lower leg and foot should have been.

Mike just looked at me with a concerned expression. We both knew the largest bandage in the kit we carried was a four by four. In this case, a four by four or a hundred of them would be like the little Dutch boy's finger trying to stop the New Orleans levee failures.

"What are we gonna do?" Mike asked.

"Watch," I told him as I undid my belt buckle, sliding the belt off my waist. I wrapped it around what was left of his lower leg, putting the tag end through the buckle and pulling as tightly as I could.

Continuing to keep the belt tight with one hand, I picked up the portable from the grass where I'd dropped it and keyed the microphone. "Lackawanna, 36-16," I said. No response. Geographically, we were in kind of a hole so I wasn't surprised by the lack of radio communication.

"Kara, dial 911 on your phone for me," I said to the boy's sister. She dialed and handed it to me while I kept a tight hold on the makeshift tourniquet.

"911, what is your emergency?" said the call taker.

"On the Snyder Road incident, do you have ALS dispatched?" I asked.

"ALS and BLS have been dispatched, sir."

"Okay, this is 36-16. I have a lower leg amputation here. Have both units expedite!" I dropped the phone and went back to work without waiting for a reply.

Mike was talking to the victim, Evan, while getting a blood pressure cuff in place on his arm. Evan was moaning, "I'm going to die."

"You are not going to die!" Mike told him.

The tourniquet had the bleeding under control but he had already lost an incredible amount of blood. Mike kept him talking to distract him and keep him conscious, while monitoring his vital signs. His blood pressure was good despite how much blood he lost.

Mike and I worked to keep both Evan and Kara as calm as possible. We needed to keep Evan out of deep shock and his sister, being with him and in control, helped. We let her get some ice and rub it on his lips. It distracted both of them while we waited for one of the rigs to arrive.

Wait, we did. The moron dispatcher never relayed the expedite request over the air, so none of the other responders knew how serious the situation was.

Evan had been working with a tractor and dump cart on a stone wall at the back of a field. The power take-off on the tractor caught his pants leg when he dismounted to check on a problem. It pulled his leg into the spinning shaft and his lower leg was history. Somehow, he had managed to get back on the tractor and drive to the house. Kara had returned home at just about the time he pulled in and collapsed off the seat of the tractor. If he hadn't made it back to the house, he would have died in that field for sure. Now we had to keep that from happening in the front yard.

We heard a vehicle approaching and Mike flagged down the medic unit with the paramedics aboard.

"His pressure is good, but we've gotta get a line into him," I told the first medic that trotted up. They started working, deciding to put a line into each arm.

Our BLS unit finally arrived along with the rescue as well as more manpower. The chief took Kara on the quad to look for the foot since Kara was the only one who understood Evan's description of where he had been working.

We got some huge trauma dressings on the stump while I continued to hold onto the belt. My hands were starting to cramp up from the strain. A back board was laid alongside Evan and we worked him onto it and then onto the stretcher. I held onto the belt with every ounce of strength I had.

Once we loaded him into the back of the ambulance, I handed the end of the belt to one of the young guys who held it all the way to the hospital. Billy, the senior EMT in the department, had arrived and jumped into the back with Mike and the two paramedics. We put a great driver in the front seat who was known for driving safe, smooth and fast, and away they went to the Trauma Center at Community Medical Center in Scranton.

A minute or so later, those of us remaining heard the quad returning. They had located the foot, and had it wrapped in ice. The chief and another guy jumped into the medic unit with the limb for a high-speed trip to the hospital.

Hours later, some news began to filter back. They weren't able to reattach the limb, but our treatment had been enough; he would live.

As for the belt, well, that stayed at the hospital. Mike told me it stayed on Evan until they took him into surgery. The trauma surgeon asked him if I wanted it back. One look told him I didn't. With the end result, I certainly didn't mind losing it that day.

• • •

The loss of Evan's leg was terrible, but winter brought even more tragedy to the area.

Two girls were huddled, soaking wet, on the snow-covered bank of the pond, their piercing screams echoing across the ice. Mack, who owned the nearby house, was splayed on his back on the ice, his face dark blue, cyanotic in the extreme. One of the first arriving responders was pumping on his chest, the CPR compressions having little effect.

Less than five minutes beforehand, the dispatcher's voice sent my heart into my throat, and a stream of adrenaline through my veins.

Report of a young girl through the ice.

Mike wasn't home, but I knew he'd be getting the same dispatch. I called his cell phone and he quickly answered.

"Go to the station. Do not even think about going to the scene," I directed.

"I'm en route to the station now," he answered.

"Good," I said and then disconnected.

I knew he wasn't far from the scene and the last thing I wanted was for him to be there first. For a young firefighter, that shot of adrenaline can, at times, override good judgment. Jumping into freezing water, child or no child, without the right equipment was a fast way to die. Emotions, particularly combined with inexperience, can overwhelm the brain.

As I headed for my car, the radio traffic from the dispatcher was becoming more confusing—not unusual for something like this. I started over the hill, thankful I would arrive before Mike.

There was a ladder laying on the ice, the tip edging out into the open water. We assumed that Mack had gotten the ladder to try to reach the little girl who must have been under the water at that point. We needed to get things organized and obtain some information on what the hell was going on. The simple secret to handling critical emergencies is to make order out of chaos.

I stood on the ice looking at the open water; the positioning of the ladder was our only clue as to where the child likely fell in. Mack would have put it at that spot.

More personnel and apparatus were arriving. The dive truck with the water and ice rescue equipment was on its way. Marty, the Dalton assistant chief, joined me on the ice.

"I've gotta give this a try," he said. I knew what he was thinking.

"Marty, this is dumb. You've got a wife and kids too. The right equipment is on the way. You can't last more than thirty seconds in there," I said.

"I know, but I've got to try," he responded.

Still disagreeing, we got him in a life vest and tied a rope onto him. I got on the end of the ladder away from the hole in the ice, and fed him the rope. As he started out, I warned him. "You've got thirty seconds, and I'm yanking your ass out of there."

He nodded in acknowledgement. I felt like shit about the kid, but I was trying to use logic over emotion.

Marty lowered himself into the icy dark water. He held his breath and ducked under to see if he could spot her. When he popped to the surface, I could see his eyes had already started to glaze over from hypothermia. I started pulling on the rope, bending his torso back onto the ice.

"Marty, get your legs up," I yelled. He did and we gained more ground. I reached out and grabbed his frozen hand and pulled him the rest of the way out of the water.

"I can't see a thing down there," he said.

"Get your ass in a warm rig and strip," I told him. No response, but he left to get out of his clothes and hopefully forestall the hypothermia that had already started to take hold.

The dive truck pulled in and seconds later, Little Billy flopped down beside me, SCUBA gear in place over his wet suit. Mike was there by that time as well and helped Billy get his fins securely fastened. He then tied a safety line off the buoyancy control vest that carried the diver's air tank and octopus. Mask in place and air regulator in his mouth, Billy slipped across the ice and into the inky blackness, his fins briefly flapping in the air as he powered his way down to begin his search. Mike monitored the rope carefully, ensuring it didn't get tangled.

Up and down Billy went numerous times, coming to the surface mainly to orient himself in the hole. He had us relocate the ladder to the opposite side of the opening while he searched that area. Visibility was next to nothing, two feet at best, he told us. After twenty minutes, far longer than he should have been in that water even with a wet suit, he came out, half collapsing on the ice. A crew helped him to the warmth of the dive truck where he could strip out of the now icy neoprene and start to warm up.

Information was beginning to firm up with reports coming back from the hospital; mainly bad news, but one positive piece. Mack had been transported and, as I had expected, didn't make it. He was pronounced dead. Then the combination of bad and good news came. We weren't looking for a little girl; she was out and okay. We were instead looking for Mack's stepson, Terry. We learned he had jumped in and gotten the little girl as well as her mother, who had also slipped in, out of the bitter water before the hypothermia became too severe. After saving them, Terry himself slipped below the surface of the water.

Luckily, Little Billy had not known who he was actually looking for. He was close friends with our—now known—victim. We had a hard enough time getting him out of the water when he thought he was looking for a stranger. If he had known Terry was down there, it would've been a war to get him out of there.

More divers were on scene now, preparing to enter the water to search for Terry. Mike worked with them, ensuring everything was zipped and fastened properly as well as tying off the safety lines. With little for me to do at this point, I walked over to the dive truck to see if Little Billy had been told the news.

When I climbed into the back, I could see he knew. He was packing and organizing equipment in the back, trying to keep busy, but there were tears on his cheeks. He turned to face me.

"Thank God, I didn't know," he said.

I grabbed him and pulled him to me, hugging him tightly. I knew what he was thinking. The words wouldn't help, but they were required.

"Billy, you did everything you could possibly do. You covered as much as you could. You were in there too long as it was," I told him.

"I thought I was looking for a little girl," he said. "Why couldn't I find him?" There were more tears now.

"Billy, you know how bad the visibility was," I answered. "Don't second guess yourself. You did everything right. Unfortunately, if you do this in a small town, inevitably something like this happens."

Old Todd from the dive team was standing there too. We both ran down a list of names that we'd known or been friends with that had died in incidents we'd responded to.

"Does your dad know?" I asked Little Billy.

"I don't think so," he responded, wiping his eyes.

"You stay here, I'll tell him."

Big Bill was running the dive rescue operation. Losing Mack, who had been a close friend, was bad enough but he'd been through that before. I walked back onto the ice to talk to him.

"Did you get the news on Mack?" I started.

"Yeah," he answered. Big Bill had expected it, too. I took a deep breath.

"Do you know who you're looking for here?"

"No," he responded. "Just a little girl."

"The girl is out and okay," I said. "You're looking for Terry. He got her out and went under."

His shoulders sagged. "No," he said as he looked at me again.

"I'm afraid so. That's the information we got back from the hospital."

"Oh God," he said his shoulders came back up. He knew he still had a job to do. I patted him on the shoulder as he walked away to brief the divers on the news that they were looking for an adult male.

More dive teams arrived, each contributing a few divers to the line up. The water was so cold they had to be carefully rotated to ensure they didn't become victims as well.

The longer the clock runs, the worse the odds get. A cold water drowning victim can be successfully revived after a longer time than someone in warm water but we were moving outside the envelope.

Alternate plans had to be made. Big Bill didn't want to hear it; he wanted the underwater camera now in use or the divers to find him. He wasn't leaving there without him, but we were running out of divers. One option was left—drain the pond.

It would take a while to set up to do that, in order to get engines and portable pumps positioned around the bank, so the divers still had some time. After this, though, it was time to pull the trigger.

There were mere minutes left before the pumping was to start. The last diver was in the water. He surfaced, and a roar went up. He had Terry. His frozen form was quickly pulled onto the ice, and then covered and loaded into a waiting ambulance. There was no chance; we all knew that.

Big Bill and Little Billy stood on the ice, holding each other, tears streaming down their faces.

"You both did everything you could do," I told them, wrapping my arms around both of them. Worthless words. We all felt helpless after the two deaths. We needed to get Big Bill checked now. The stress of the situation and his past medical history—he'd previously undergone by-pass surgery—had many of us worried all day.

The medics got him on a monitor, and his rhythm was okay, but his blood pressure was dangerously high. It started to come down as he relaxed, though.

The little girl had made it—Mack's stepson having sacrificed himself to save her. It was, I guess, something good to take away.

Still, on days like this, I hate this job.

• • •

But, we keep on going to calls, and the learning never ends. With all he had experienced, what Mike had been working toward for four years still awaited him. His eighteenth birthday came and went, an important transition into adulthood, but not with the excitement he would have liked. Mike turned eighteen with the hope of an immediate string of working fires since he was finally able to go inside for the first time. It didn't happen like that. In fact, it was overly quiet. As the weeks went by, his level of frustration, and the associated bitching, increased. I understood, remembering a similar wait after I turned eighteen.

Shortly after midnight on a moderate January morning, the pager went off for a chimney fire right around the corner from our house. Mike came down the hall toward our bedroom to see if I was up. Initially I tried to dissuade him. I was whipped. We had been out the night before on a surround and drown warehouse fire; the first actual fire in months but no real work other than the aerial ladders and tankers hauling the water. The lack of sleep, however, had taken its toll and I had an important meeting in the morning.

But Mike was not dissuaded. So grumbling all the way about going to a routine chimney fire, I stomped downstairs, put my bunker pants on, and met my now impatient son in the car. We made the short drive around the corner and parked just past the handsome white farmhouse. Putting on the balance of our gear, we walked up to the back of the house to check on the conditions. An elderly lady in a wheelchair was at the bottom of a ramp leading from the rear door. We walked in the back door to find three men still in the house, the first floor charged with smoke, the other side of the kitchen just barely visible. So much for the routine chimney fire.

"Everybody out now," I yelled into the grayish smoke. The three men exited and confirmed that no one else was in the house. A quick check of the basement from the outside Bilco doors verified there was no fire down there, only light smoke.

The engine arrived, and Mike pulled an inch and three quarter hose line to the front door while I put on a pack. Mike put a pack on as well and we pulled our face pieces over our heads and entered the living room. A quick check of the wall that the chimney was on showed it to be cool, surprising for the amount of smoke present. The same was found upstairs. There was too much smoke for this to be just a chimney fire, but that was where we had to start.

Shortly after we started work on emptying the wood stove, the fire decided to show itself. Flames blew out of a small ventilation fan in the wall separating the living room and kitchen. Mike went for the hose line at the front door, and I hit the flames licking out of the vent with a small stream from the pressurized water extinguisher we had brought in.

Returning with the nozzle, Mike waited for water. After about thirty seconds, the line came to life, fat and hard, and Mike opened the nozzle on straight stream right into the still flaming vent. I knew if we had fire in this separating wall, so far from the chimney, that it was spreading in the concealed spaces. It didn't stay concealed long.

"Kitchen's off!" Jerry, the assistant chief, hollered from the doorway between the living room and kitchen.

We moved the line into the kitchen, making the hard right turn. Fire rolled across the ceiling above the kitchen cupboards. I pulled more line into the room and yelled through my face piece, "Get it, Mike!"

He braced himself on the floor and racked the bail back, the straight stream again blasting from the black nozzle tip, pounding the fire quickly pushing it back. A few seconds of water and the flames were gone. Mike shut the nozzle and we looked, as best as possible, through the smoke at where the fire had been.

"Hit that hole up there," I told him. I was sure there was more fire in the wall and ceiling. If we got some water in there while waiting for more manpower and tools perhaps we could slow it down. The straight stream again thumped the junction of the wall and ceiling. After about fifteen seconds of water, I had him shut it down again.

Now we could see smoke pushing from behind a high cabinet in the corner. I crawled up to it and pulled the door open. Flames blew out of the cabinet into the room and up the wall. Mike heaved on the line and got around the now open cabinet door and killed those flames as well. Jerry handed me a pike pole and I smashed the kitchen window next to the cabinet. Smoke was to the floor by this point, our visibility poor.

"Vent out the window, Mike!" I yelled.

He opened the nozzle on a narrow fog, something he had practiced innumerable times in training, but had never done for real. He watched the smoke become entrained with the flowing water, streaming out the opening together.

From beneath his face piece came the first words I'd heard him utter since we entered the building—"Aw, cool."

There was no further fire evident in the kitchen, so we pulled the line back to the living room. Hook still in hand, a few quick pulls of the ceiling revealed fire running in the channels of the heavy floor joists above. An eighty square foot area was well involved. A few more shots with the hose line and we were in good shape.

My low air alarm began sounding. Mike still had plenty of air left, which I heard about in great detail in the days that followed, but I didn't know any of the additional people now working the room. He had come in with me so he would leave with me.

"Lets go, I'm low on air," I said.

He followed me out the front door and onto the snow covered lawn. We both knelt down and removed our helmets and face pieces.

My son and I, together.

I looked over at him as he stared at the house, now only light smoke was coming from the top of the front door.

"Did you enjoy it?" I asked.

He just nodded back at me, a satisfied look on his face.

I always thought it would be great to be there for his first time inside, but I never knew if it would actually happen. I had just lived a dream. Emotions welled up inside me, flowing through my system. I thought I was going to cry. He wasn't my little boy anymore.

• EPILOGUE •

Change is the Only Constant

In my mid forties, twin digital hearing aids became my near constant companions. These new friends were introduced when my hearing loss was diagnosed at fifty percent, allowing me to accurately claim I was "half deaf." My hearing loss was noted much earlier and I used a single, less sophisticated aid for many years until I got the digital ones. I am sure standing up in jump seats, in close proximity to Federal Q sirens and twin air horns, didn't help matters. Working with loud gas operated power tools also damaged my hearing. That said, the end result was of my own making. The good news is that the enclosed cabs of current and new apparatus will help reduce future firefighters'—Mike's in particular—exposure to such noise levels.

I probably would have "retired" by now if it was not for Mike. I've enjoyed the calls we've run together and, if I get a few more, even better. I will be happy to fade away, content with my experiences and contributions. I will never be one of those guys that just drive or direct traffic once they decide not to do interior work any longer. That would frustrate me to no end. When I stop fighting fires, that will be the end.

The fire service has changed immensely in the last thirty years. One major factor was 9/11, but it just accelerated a process already underway in emergency services. That term, emergency services, is probably more representative of what the fire department has actually become—a do everything and go to agency for emergencies. Actual fires are now one of the smallest percentages of incidents that we respond to.

In many communities, emergency medical services (EMS) are now operated by the fire department; add to that hazardous materials, confined space rescue, swift water rescue, dive rescue, trench rescue, high angle rescue, urban search and rescue (USAR), potential response to terrorism incidents (such as anthrax, etc.), code enforcement, fire prevention, etc. If a new hazard requiring trained personnel emerges—the fire department will be expected to handle it. They've got all those big trucks and fancy equipment. They'll probably know how to deal with it.

It's to the point where I'm very glad to be at the tail end of my "career." I'm one of the minority feeling we've gone "a bridge too far" when it comes to all of these new responsibilities. Some fire departments are actually doing bomb squad work, for

Christ's sake. The amount of training to become proficient in all this stuff—never mind maintain a level of proficiency long term—has become overwhelming, particularly for volunteer services. It reminds me of the old saying—*jack of all trades, master of none*. The problem is, if you don't master this stuff, people can die and some of those people may be firefighters. Bottom line—this is no longer the job I signed up for.

It's interesting, watching Mike develop in this era of increased training and certification requirements, the mind set changing on how things should be done. The more I watch things change, the older I feel. I used to be on the cutting edge. Now I look at things with a jaundiced eye. That doesn't mean I've become a dinosaur; I'll quit completely before that happens. But I have lost the illusion that all change is good.

There are some very positive things going on in fire service today. An overall increased emphasis on safety is a good thing. "Everyone goes home alive" is a commendable motto. However, I don't always agree with how we are achieving these goals.

Training young firefighters in artificial environments in burn buildings with pallets and hay, gas jet fires, and no actual fire spread is a great way to build confidence. It's also a great way to build young firefighters that don't have the faintest idea what a real fire is like; how fast it can spread, and how quickly it can get behind you. That's how you get in trouble.

We do a much better job protecting our people today with high quality fire-resistant protective clothing. But, again, there's a downside. It makes it much easier to go farther into a fire, too far in, where it's easy to get in trouble. I certainly don't advocate a return to three quarter boots and no hoods, but at least our bodies told us how far was too far when wearing that equipment. Now we must utilize judgment, which is tough to do if you don't have any.

Firefighter survival classes are great, as far as they go. We teach breaking through walls, sliding ladders head first, bail-ing out windows, and similar techniques. As one who has actually slid a ladder head first in a critical life-threatening situation, I have news for you: if you need to do it, it comes naturally. No special instructions are needed. How about spending a bit more time on size-up and situational awareness? Not getting in trouble in the first place might be a bit more desirable.

I love those infomercials for those wrenches that do everything; one socket fits every nut, a single wrench for every application. Ever see the mechanic that works on your car use one of those silly-ass things? That's what our do everything pieces—Quints, pumper-tankers, rescue-engines—remind me of. They may have the capability to perform more than one function, but do they do any of them well?

Your Quint pulls in first and is being operated as the attack engine on side A. The ladder is suddenly needed on side B or C. Oops, oh well, can't be moved, since we have attack lines off. I can think of similar analogies for our pumper-tanker or any other do everything piece you can name.

The argument I hear is that we don't have the manpower to crew two pieces properly. The last I knew—if we ran a Quint in as the first due piece, it didn't change the fact that we still needed to accomplish both engine and truck company functions to handle the fire properly. If we don't have the manpower for an engine and a truck, regardless of the fact that we have this fancy Quint, how are we going to get the work done?

• • •

When I was a rookie in upstate New York, the old guys were those that had made interior attacks with either filter masks or no mask at all. When my father started, there were only two air packs for the whole department.

In Maryland, the old guy/new guy dividing line was the anti-war riots that took place on the University of Maryland campus during the Vietnam War. If you worked the riots, you were an old-timer. If not, you weren't.

Now the dividing line is whether or not you ever rode the back step of an engine or wore three-quarter boots to a fire. Departments left both behind many years ago due to safety concerns. Soon the line will shift again. It may be open jump seats versus fully enclosed cabs. It might be something else. But no matter what, the line will shift again. If there is a fourth generation to continue the "family business," the dividing line will always be there.

CPSIA information can be obtained at www.ICGtesting.com
Printed in the USA
LVOW111548140212

268669LV00003B/8/P

9 780982 256596